GRADE 4

Discussion Book

Author

Calvin Irons

Contributing Authors

Sandra L. Atkins

Claire Owen

Authors Calvin J. Irons, Ph.D.
Thomas E. Rowan, Ph.D.

Consultants

Laurie Godwin
Assistant Principal
Aurora Public Schools, Aurora, CO

Eileen Harris, Ph.D.
Director of Planning and Evaluation
Charlotte County School District, Port Charlotte, FL

Pat Huellmantel
K–12 Math/Science Coordinator
Flint Community Schools, Flint, MI

Mari Muri
Mathematics Consultant
Connecticut State Department of Education, Hartford, CT

Barbara Ramsey
Curriculum Resource Teacher – Math and Science
Santee School District, Santee, CA

Michelle Rohr
Director of Mathematics
Houston Independent School District, Houston, TX

Cover Illustration Susan Swan

Illustrations Kelvin Hawley: p. 67; Dianne Hill: pp. 14, 30, 31, 44, 49, 71, 73, 79, 88, 92

Photography Page 6 (lobelia) © Royalty-Free/CORBIS; page 8 (divers) © Duomo/CORBIS; page 9 (runners/female) © Index Stock/Zefa Visual Media; page 15 (paint box) © David Young-Wolff/PhotoEdit, Inc.; page 16 (football stadium crowd) © Mary Steinbacher/PhotoEdit, Inc.; page 21 (girl with compass) © Roger Ressmeyer/CORBIS; page 32 (car factory) © Charles O'Rear/CORBIS; page 42 (dinosaur skeleton) © Rudy Von Briel/PhotoEdit, Inc.; page 63 (athlete throwing shot-put) © Chris Hamilton/CORBIS; page 68 (colonists) © Bettmann/CORBIS; page 77 (two boys and basketball hoop) © Ty Allison/Getty Images; page 78 (running marathon) © John Neubauer/PhotoEdit, Inc.

Mike Fisher: pp. 86, 87; Fotosearch: p. 6 (petunias); Getty Images: pp. 6, 13, 15, 19, 34, 45, 47, 48, 83, 93; Gordon Hill: pp. 20, 29; The Image Bank; pp. 20, 23; David Johns: pp.10, 20–23, 27, 33, 35, 37, 38, 43, 44, 52, 59, 60, 62, 66, 69, 76, 82, 84, 85, 89, 91, 92, 94, 96; Jeff Richey: pp. 24, 70, 79, 80, 90

Acknowledgments The authors are grateful to all the teachers and students who participated in the development of this project. Special thanks go to the following teachers and their students:

Laurie Godwin, Tollgate Elementary School, Aurora, Colorado; Duncan Rasmussen, Helen Cera, and students at Penleigh and Essendon Elementary School, Melbourne; Cherie Brown, Fitzroy Elementary School, Melbourne; Oak View Elementary School, Maryland; Olney Elementary School, Maryland; Broad Acres Elementary School, Maryland; Dufief Elementary School, Maryland.

Thanks also go to Lara Whitehead, Hour Bak, Jenny Randle, Emily Johns, and Nicole Polidoras.

www.WrightGroup.com

Printed in the United States of America.

Send all inquiries to:
Wright Group/McGraw-Hill
P.O. Box 812960
Chicago, IL 60681

ISBN 978-1-4045-6579-1
MHID 1-4045-6579-5

2 3 4 5 6 7 8 9 VHF 13 12 11 10 09 08

The McGraw·Hill Companies

CONTENTS

1. Describe some ways a landscape gardener might use mathematics.
2. What geometric shapes do you see in this picture?
3. Where do you see measurement?
4. What other mathematics could be used here?

1. What geometric shapes can you find in this plan? List them.
2. a. Look at the tiles. Which shapes would fit together without leaving any gaps? List them.
   b. Do you know any other shapes that *tessellate*?
3. Look at lines drawn on the plan. Point to an example of
   - a diagonal line
   - a straight line
   - a curved line
   - parallel lines
   - intersecting lines
4. What kinds of angles can you find in the plan? List them.
5. What mathematics would you need to draw a garden plan?

1. Which flower requires

a. the most space between each plant?

b. the least space between each plant?

2. **a.** How many lobelia plants would you buy for a row that is 6 feet long?

b. How much money would you spend?

3. Suppose you needed 15 plants. What would be the total cost if

a. all the plants were impatiens?

b. 10 plants were petunias and 5 were snapdragons?

4. A rectangular garden bed is 15 inches by 3 feet. Which plants would you buy? How much would you spend?

EVERGREEN

Price per plant	
Lobelia	49¢
Impatiens	75¢
Marigold	85¢
Petunia	59¢
Snapdragon	65¢

1. The Evergreen Nursery keeps records of plants sold in pots. What does each graph tell you?
2. Which month showed:
   a. the smallest number of plants sold?
   b. the lowest sales value?
3. Why do you think your answers for Question 2 are not the same?
4. Look at the month of September.
   a. How many plants were sold?
   b. What was the sales value?
   c. About how much did each plant cost?
5. What was the total number of plants sold in the 6 months?
6. About what fraction of the total number of plants was sold in November and December?
7. If you were the owner of the Evergreen Nursery, what other information would you want to have?

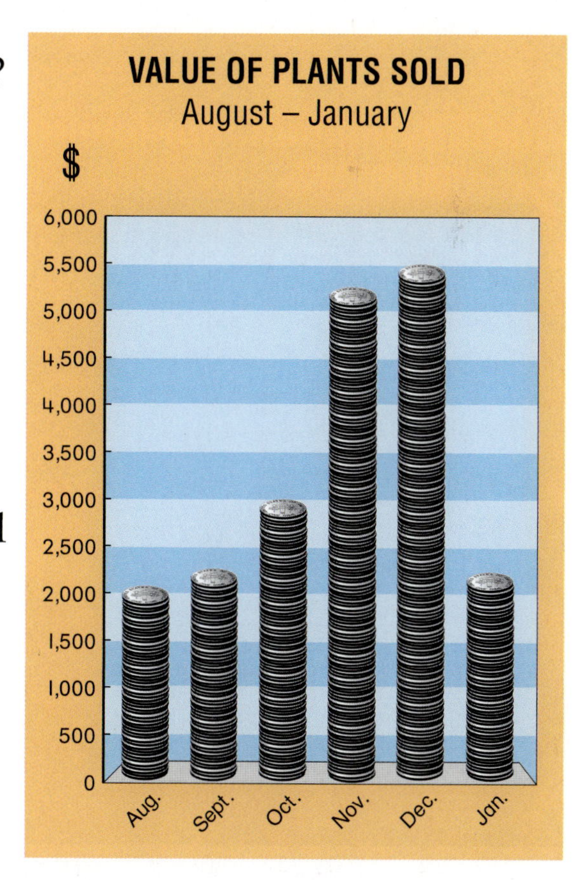

OLYMPIC SWIMMING EVENTS

1. Which swimming event is closest in length to 1 kilometer?

The length of an Olympic swimming pool is 50 meters.

2. For each distance below, how many pool lengths do you swim?
   a. 100 meters
   b. 200 meters
   c. 400 meters
   How do you know?

3. How many pool lengths are there in these relay events?
   a. 4 × 100 meters
   b. 4 × 200 meters

4. How many kilometers long is the 1,500 meter event? Could you say it another way?

5. How much less than 1 kilometer is each relay event?

OLYMPIC TRACK AND ROAD EVENTS

1. Which track event is closest in length to 1 kilometer?
2. Which track events are less than 1 kilometer?
   How much less than 1 kilometer is each of these events?
3. One lap of a running track is 400 meters.
   How many laps would you run in these events?
   a. 1,500 meters **b.** 5,000 meters **c.** 10,000 meters
   What are some different ways you could figure out each answer?
4. The marathon is a little more than twice the length of the 20,000 meter road walk. What might the length of the marathon be?

Paper Clip Facts

1. Estimate how many paper clips you need to make a 1-meter chain.

2. Stretch out a paper clip and estimate its length.

   How many paper clips do you think could be made from a
   a. 1-meter roll?
   b. 10-meter roll?
   c. 25-meter roll?

3. How would you describe the thickness of a paper clip?

> The millimeter is the smallest metric unit of length used in everyday activities. A paper clip is about 1 millimeter thick.

4. Make a list of some other things that are about 1 millimeter in thickness or width.

5. What metric unit would you use to weigh a paper clip? What do you think a paper clip might weigh?

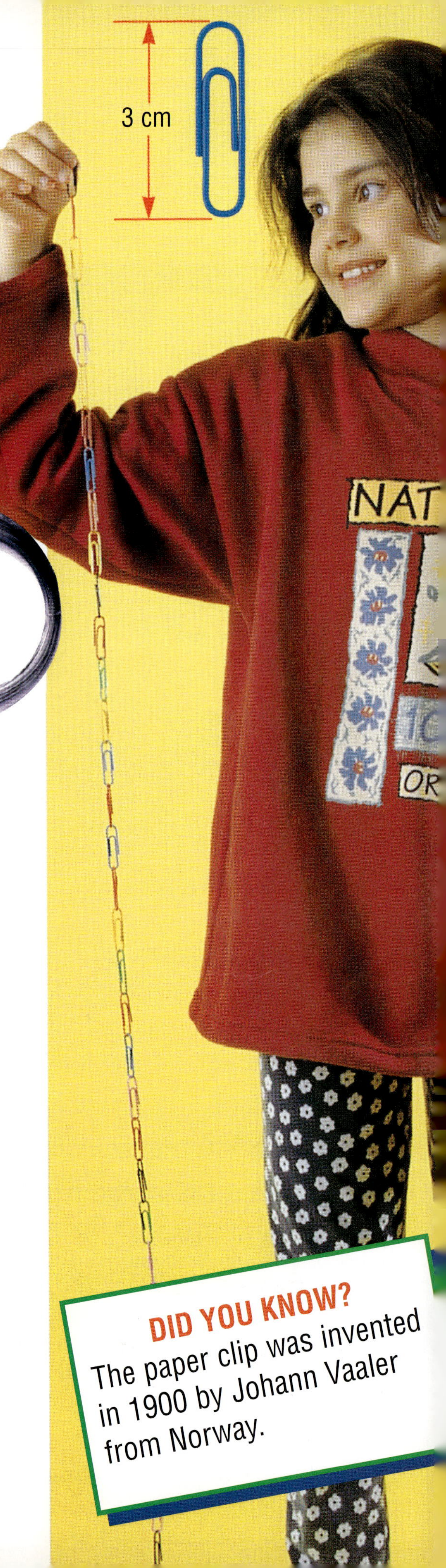

DID YOU KNOW?
The paper clip was invented in 1900 by Johann Vaaler from Norway.

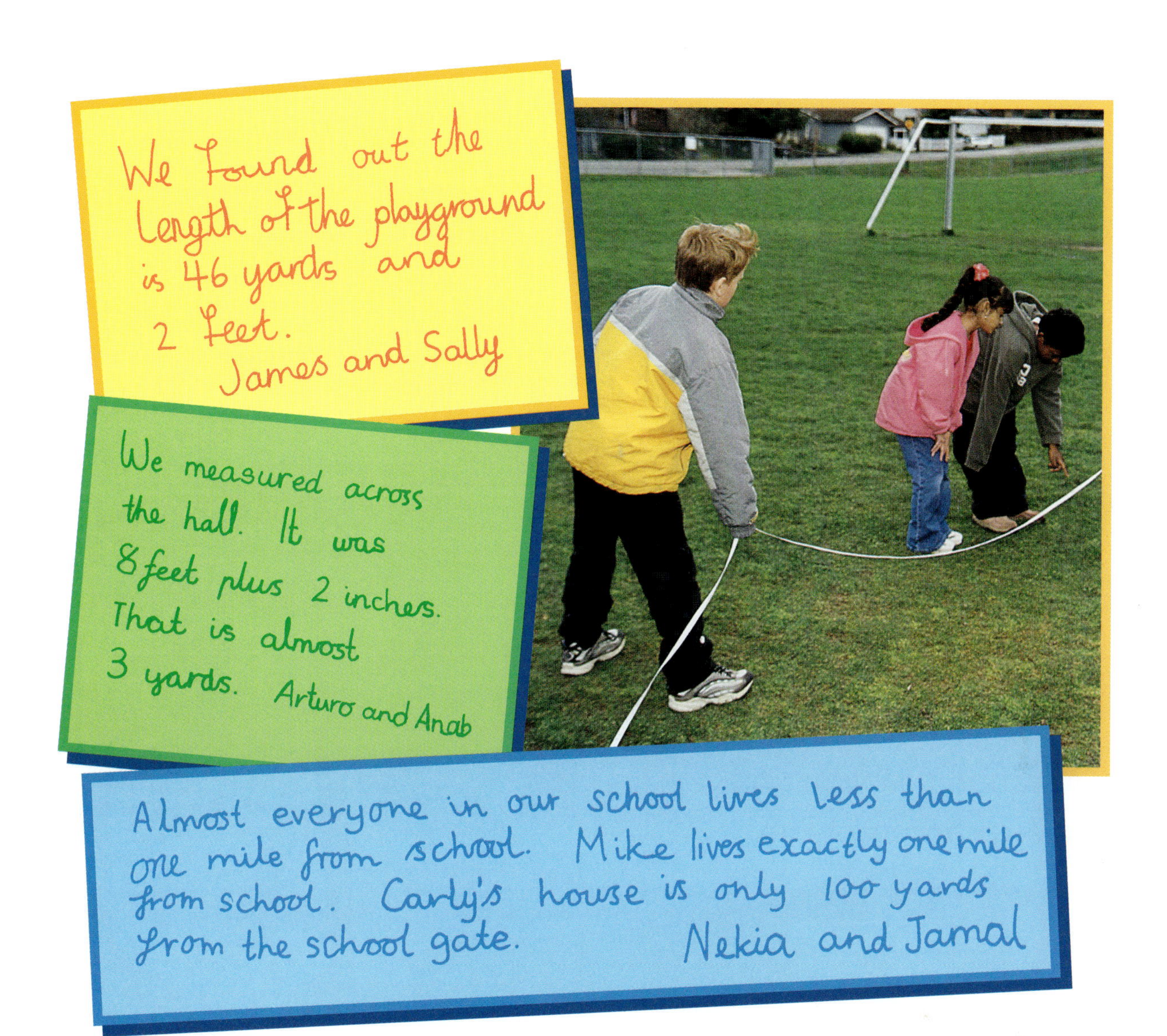

1. James and Sally measured yards and feet. What distance would they have written if they had measured in
   a. feet? **b.** inches?
2. Arturo and Anab measured in feet and inches. What are some distances you would measure in feet and inches?
3. What are some distances you would measure in yards but not inches?
4. Nekia and Jamal found out that 1 mile is close to 5,000 feet. About how many yards is that? How would you figure it out?

1. How far is it from New York to San Francisco along Interstate 80? How did you find out?

2. What is the distance between these cities?
   a. New York and Cleveland
   b. New York and Salt Lake City
   c. Salt Lake City and Sacramento

   Explain how you calculated each distance.

3. Use the distances on the signpost to make up an addition problem and a subtraction problem.

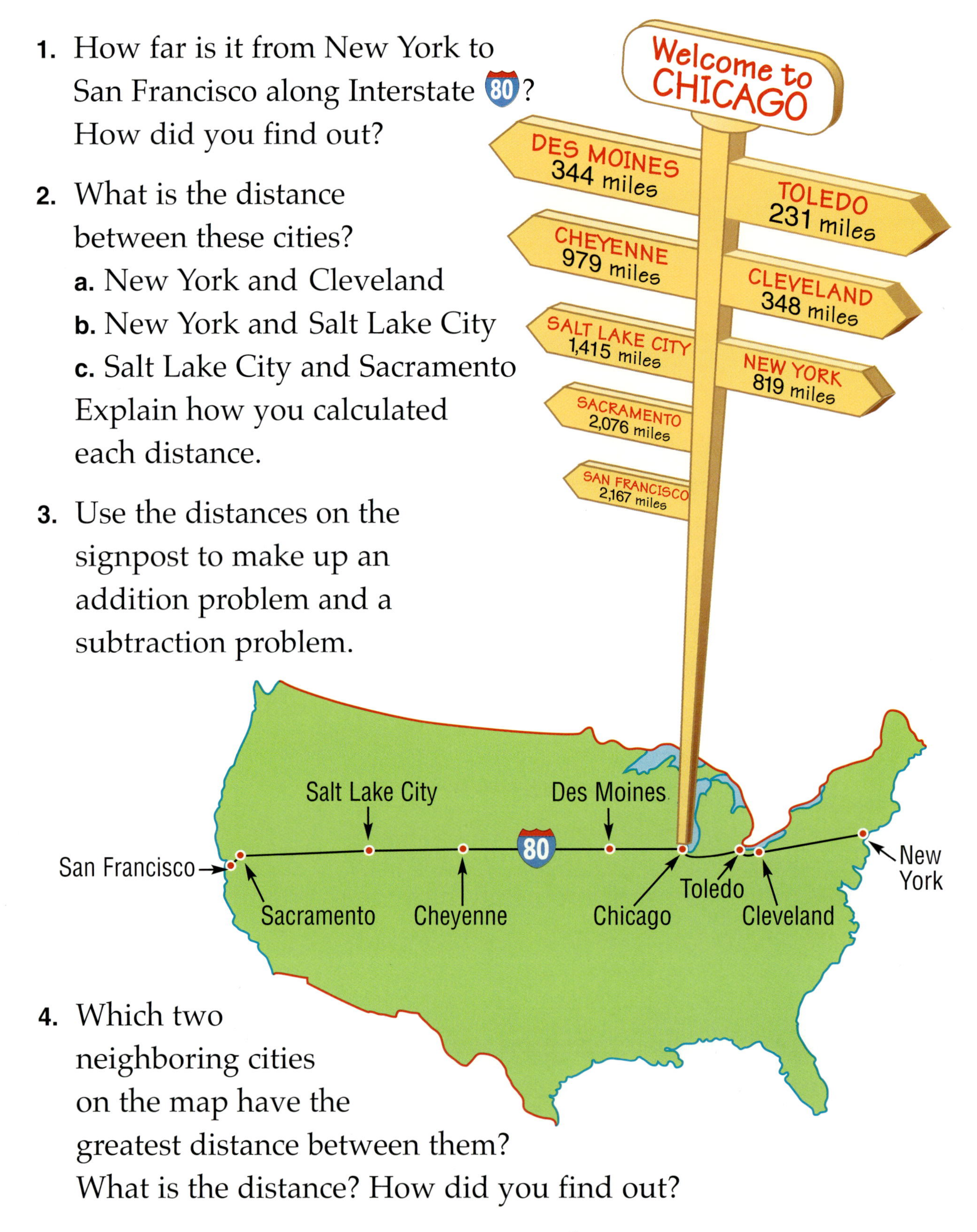

4. Which two neighboring cities on the map have the greatest distance between them? What is the distance? How did you find out?

5. If you travel 1,200 miles west from Chicago, how much farther is Salt Lake City?

1. If you were going to be in San Francisco for 7 nights, which hotel would you choose? How much would you pay?
2. How much cheaper is it to stay 7 nights at the Chicago Oak Brook than 7 nights at Chicago O'Hare? How did you figure it out?
3. How much more expensive is 7 nights at the New York Manhattan than 7 nights at the Des Moines City?
4. The regular daily rate at Chicago O'Hare is $139. How much do you save when you stay for 7 nights and pay the summer special rates?
5. Use the 80 map to plan a trip from Chicago to New York and back. Plan where you will stay each night. Estimate the total cost of accommodation. Share your plan with the class.

1. What would you pay for these orders at Sam's?

a.
White bread
Chicken
Tomato

b.
Half Sub
Ham and Pastrami
The Works

c.

2. If you had $2 to spend on sandwiches, what would you order? What change would you get?
3. If you were buying sandwiches for four people at Sam's, how much money do you think you would need to take? Explain your thinking.
4. How much would you take to pay for sandwiches for **your** family at Sam's?
5. How could you spend exactly $8 at Sam's if you bought one sandwich each day for 5 days?

1. What did these children buy? 

Josie	Carmetta
I bought two different types of supplies. I spent $2.28.	I bought three different types of supplies. I spent $2.38.

Ling	Tony
I bought three different types of supplies. I spent $2.77.	I bought four different types of supplies. I spent $2.26.

2. What supplies can Jamie buy if she spends more than $3.00 but less than $3.50? Explain how you figured out your answer. 

3. Make up your own questions about the supplies in the picture. Exchange your questions with a partner.

Sally gave these examples of large numbers.

One thousand	Ten thousand
• number of children in my school	• number of people in my town
• number of seats in our auditorium	• number of seats at the football stadium
• number of magazines sold at the bookstore in one week	• number of newspapers sold in town in one week
• number of drinks sold at our school in one week	• number of drinks sold at our school in ten weeks

1. Work as a class to list some things you think of as **one thousand** and **ten thousand**. Try to choose things that can be checked. This is what Sally discovered when she checked the examples above.

• 876 children at my school	• 24,367 people in our town
• 1,150 seats in the auditorium	• 14,250 seats at the football stadium
• 2,635 magazines sold	• 36,296 newspapers sold
• 1,724 drinks sold	• 21,306 drinks sold

2. Compare Sally's two lists. What do you find surprising?

3. Work in groups to check the estimates on your class list. Compare the actual numbers with the estimates.

1. Sally wrote the population of her town on an expander. How do you say the population of Sally's town?

Then Sally closed part of the expander. How does the partly closed expander help you read and say the number?

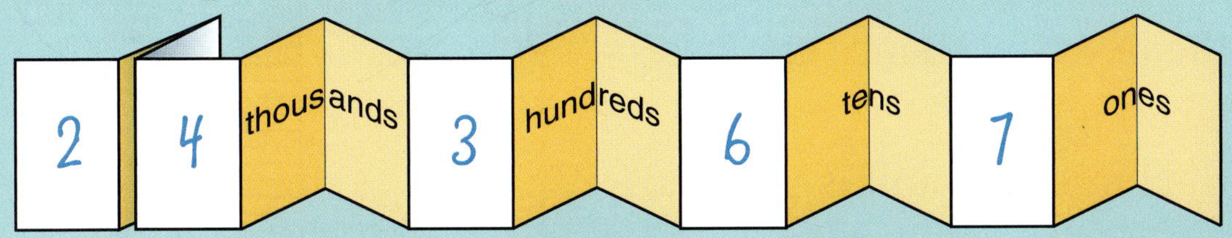

2. Sally wrote some of her other large numbers on expanders. Read each number aloud.

3. Working in groups, experiment with these digits on a number expander.

2 6 3 8 1

List all the numbers you were able to make.
Write 6 of the numbers in words.

1. What is the target amount?
2. How much money has been raised so far?
3. How much more money is needed before
   a. the hospital beds can be bought?
   b. the new ambulance can be bought?
   c. the target amount is reached? Explain your answers.
4. At what target levels can the hospital buy
   a. the whirlpool?
   b. the exercise equipment?
   c. the infant crib?
   d. the new ambulance?
5. It has taken the town 10 weeks to raise $32,000.
   a. About how much has been given each week?
   b. How much longer do you think it will take the town to reach the target amount?
6. If only $52,000 was raised, and the hospital beds had been bought, how would you spend the rest of the money?

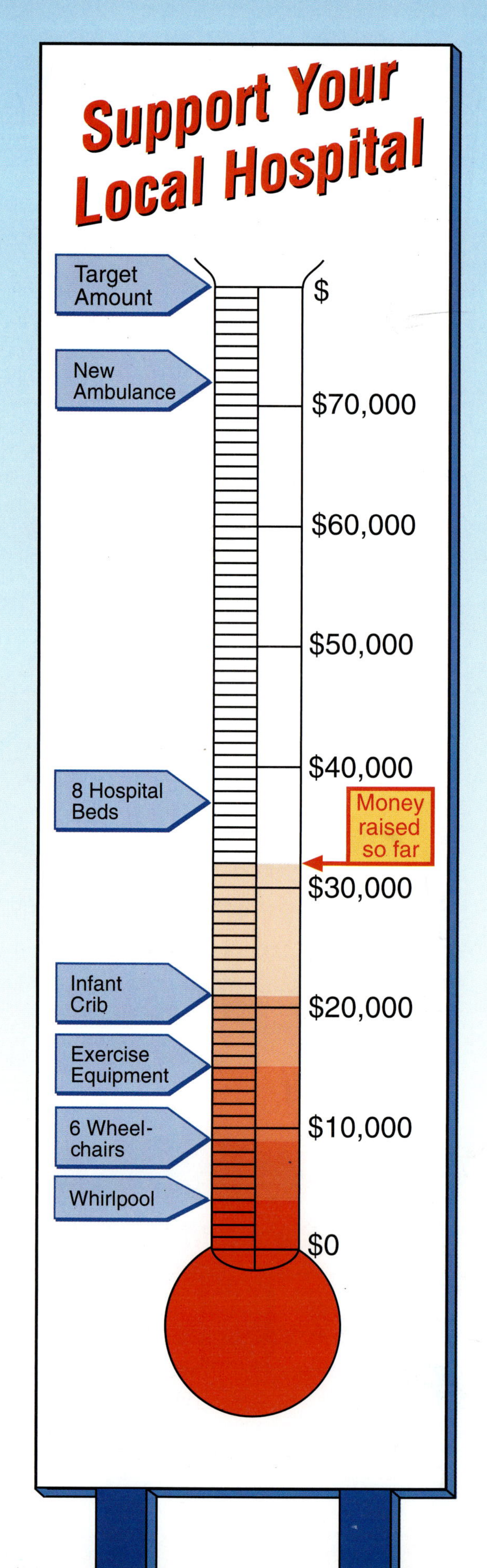

The Great Lakes

1. Which lake has the greatest area?
2. Which lake has an area that is closest to 10,000 square miles?
3. Which lake has the greater area:
   a. Lake Ontario or Lake Erie?
   b. Lake Michigan or Lake Huron?
4. Anita compared Lake Michigan with Lake Ontario and saw that Lake Michigan is larger. About how much larger is it?
5. Compare Lake Erie with Lake Superior. Try to give a mathematical comparison.
6. Which two lakes have a combined area that is about the same as Lake Superior?

1. Use these steps to make a pinwheel.

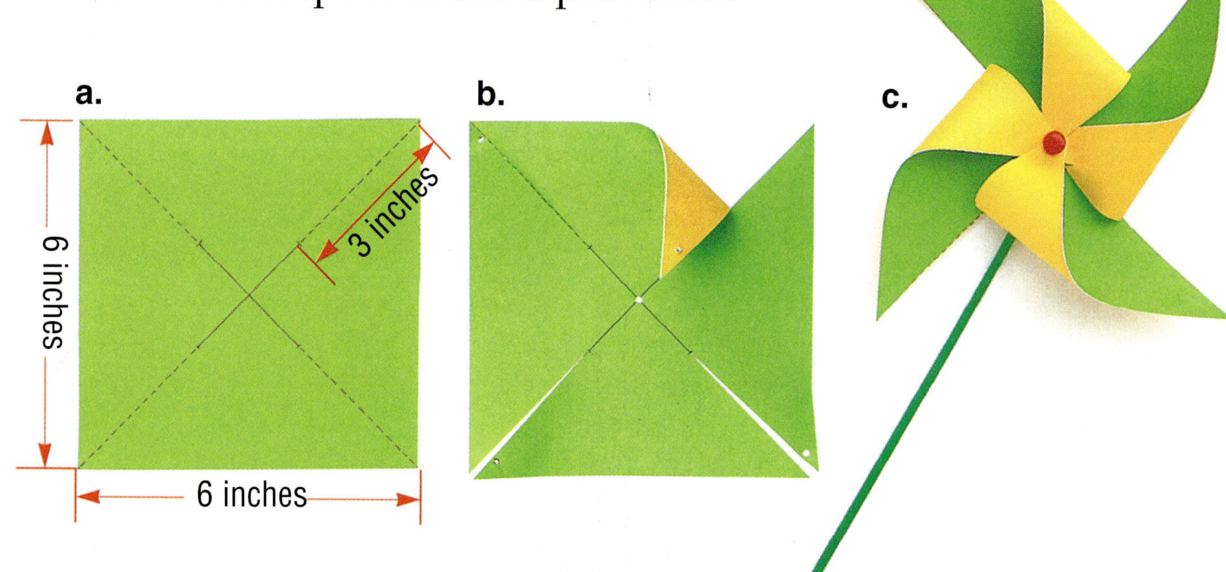

Give your finished pinwheel a quarter-turn. What do you notice about the shapes?

2. Look at each picture below. Describe how each object shows symmetry when it turns.

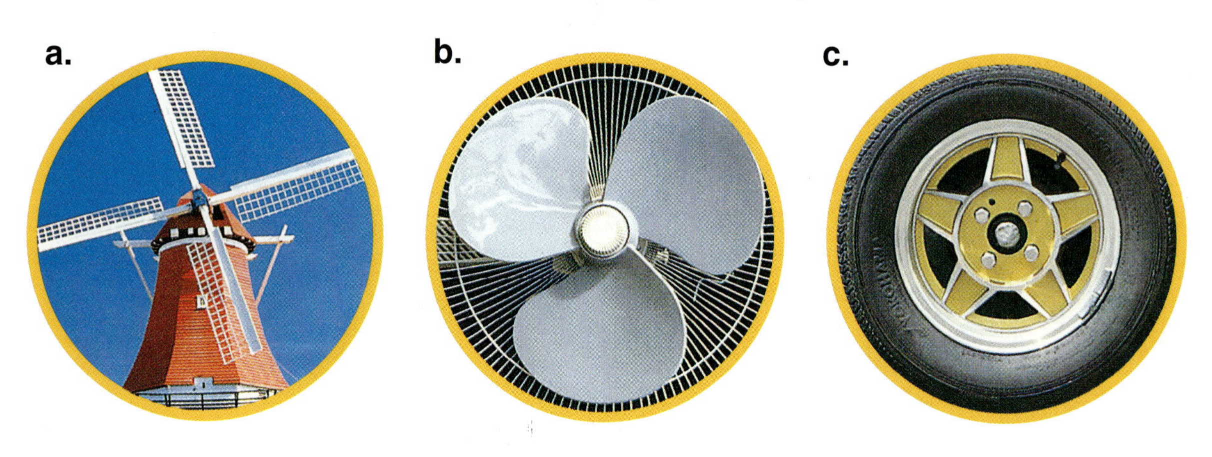

3. Copy this design onto grid paper. Turn it through four quarter-turns, and describe what you see.

4. Create a design that will look the same each time you give it a quarter-turn. Use grid paper to help.

1. What do the letters on this compass tell you?
2. The man on the compass is facing north. How would you direct him to face
   a. east? **b.** west? **c.** south?
3. If he was facing southwest, what directions would you give him to face
   a. northeast? **b.** northwest?
4. Look at the other compass.
   a. How could you use it to give directions for making turns?
   b. Use it to find another way to describe northeast.
5. Look at the row of clocks. Describe how the distance between the clock's hands changes each time.

6. List some times when
   a. the hands on clocks are closest together.
   b. the hands on clocks are farthest apart.

Clinton used these strips to make three polygons.

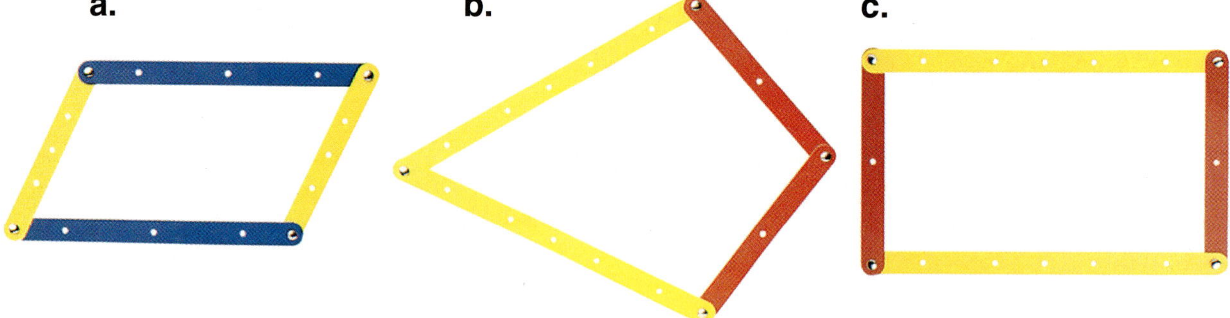

1. Look at each polygon in turn. Describe what is the same and what is different about
   a. its sides.
   b. the **angles** made at its corners.

2. Which polygon has
   a. four angles that are the same?
   b. two pairs of angles that are the same?
   c. one pair of angles that are the same?

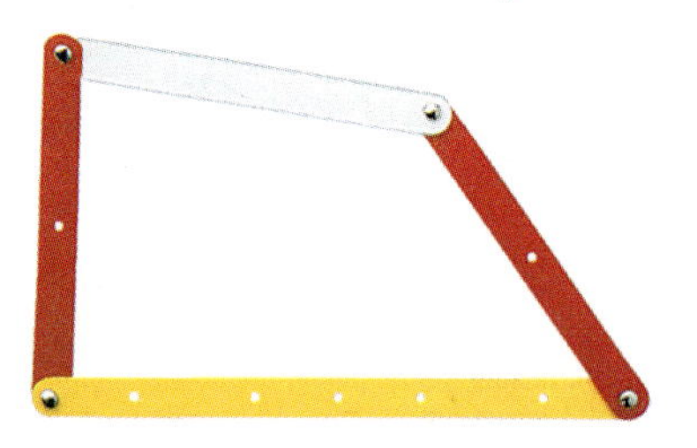

3. Ana made this polygon. At which corner are the "arms" farthest apart? At which corner are the "arms" closest together?

4. Use strips to make Ana's polygon. Change it to a new polygon and sketch it.

5. Sketch at least two polygons you could make with each set of strips shown.

a.

b.

c.

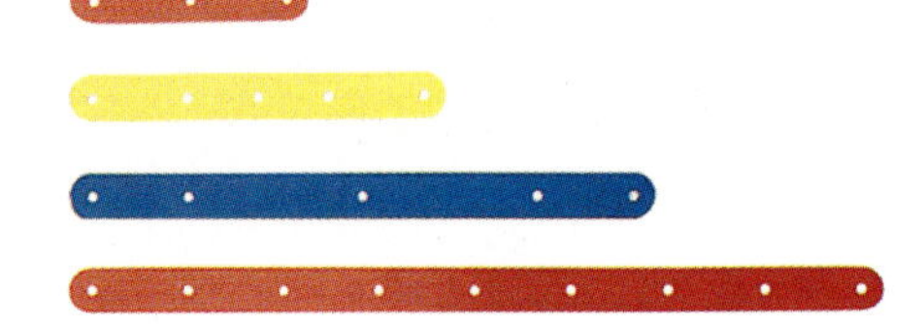

1. What polygons can you see in the picture of this bridge? Which polygon is used the most?
2. Ana's polygon was floppy. To make it rigid, she added one strip.
   a. How did she do it?
   b. What new polygons do you see?
3. Look at the shapes in the chart. What strips would you add to make each shape more rigid? Use the smallest number possible.
4. Copy and complete a chart like the one shown. What patterns do you see?
5. Predict the smallest number of strips you would need to make a ten-sided shape rigid.

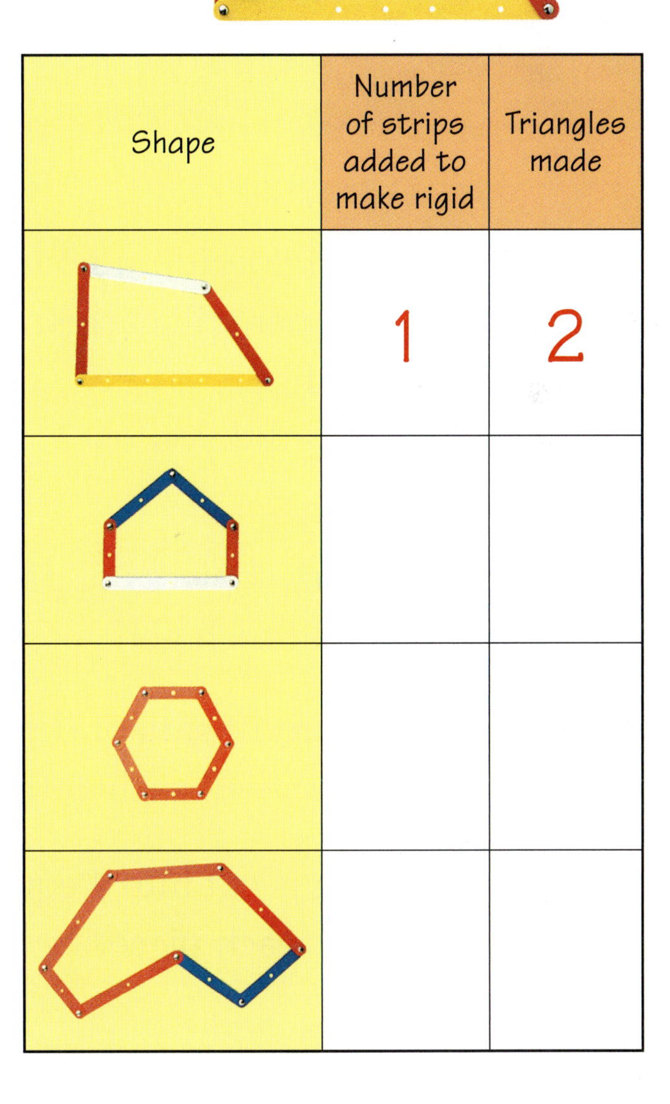

Shape	Number of strips added to make rigid	Triangles made
	1	2

1. Eight students are deciding how to share 12 pieces of paper.
   a. How much paper will each student get?
   b. How did you figure it out?
   c. What numbers did you use?

2. Solve these problems:
   a. 3 students share 4 pieces of paper.
   How much will each get?

   b. 8 students share 10 bars of clay.
   How many bars will each get?

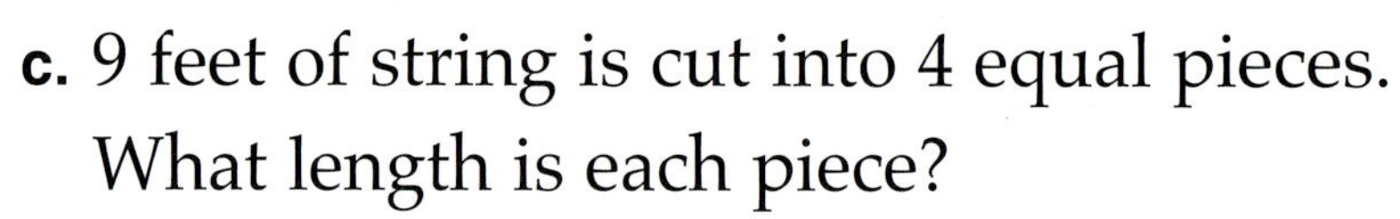

   c. 9 feet of string is cut into 4 equal pieces.
   What length is each piece?

1. Here are some problems Grade 4 students wrote.
How would you find each answer?

8 students on a team
were given 14 hot dogs.
What share did each get?

6 friends have 2 cartons
of juice to share. How
much will each get?

6 people. 4 buckets of
popcorn. How much popcorn
in each person's share?

2. Answer the questions about the eggs.

4 people share
5 dozen eggs.

a. How many full cartons of eggs does each person get?
b. What fraction of the extra carton does each person get?
c. How many eggs does each person get?

3. Write a problem that has a fraction as the answer.

Look at each Pizza Party table.

1. At which table would a person get a greater amount of pizza?
   a. Vegetarian or Hawaiian Pizza table
   b. Hawaiian or Pepperoni Pizza table
   c. Pepperoni or Deluxe Pizza table

2. What fraction of the pizza will each person get at the
   a. Vegetarian Pizza table?
   b. Hawaiian Pizza table?
   c. Pepperoni Pizza table?
   d. Deluxe Pizza table?

3. Make up a real-world problem that compares two fractions. Solve it yourself, and then give it to a friend. Did you both solve it the same way?

Some students traced around 2 yellow Pattern Blocks. They used the shape they drew to show different fractions.

1. How much of the shape is covered by
   a. one yellow block?
   b. one red block?
   c. one blue block?
   d. one green block?

2. What are four different ways you could cover one-half of the shape? Give the fractions.

3. Give two different ways to cover one-fourth of the shape. Explain your thinking.

4. Draw an outline around 4 yellow Pattern Blocks. Find different ways to cover the shape you made. Each time, use blocks of only one color. Write the fractions that describe how the different kinds of blocks covered **one-fourth** of the shape.

Stars and Stripes

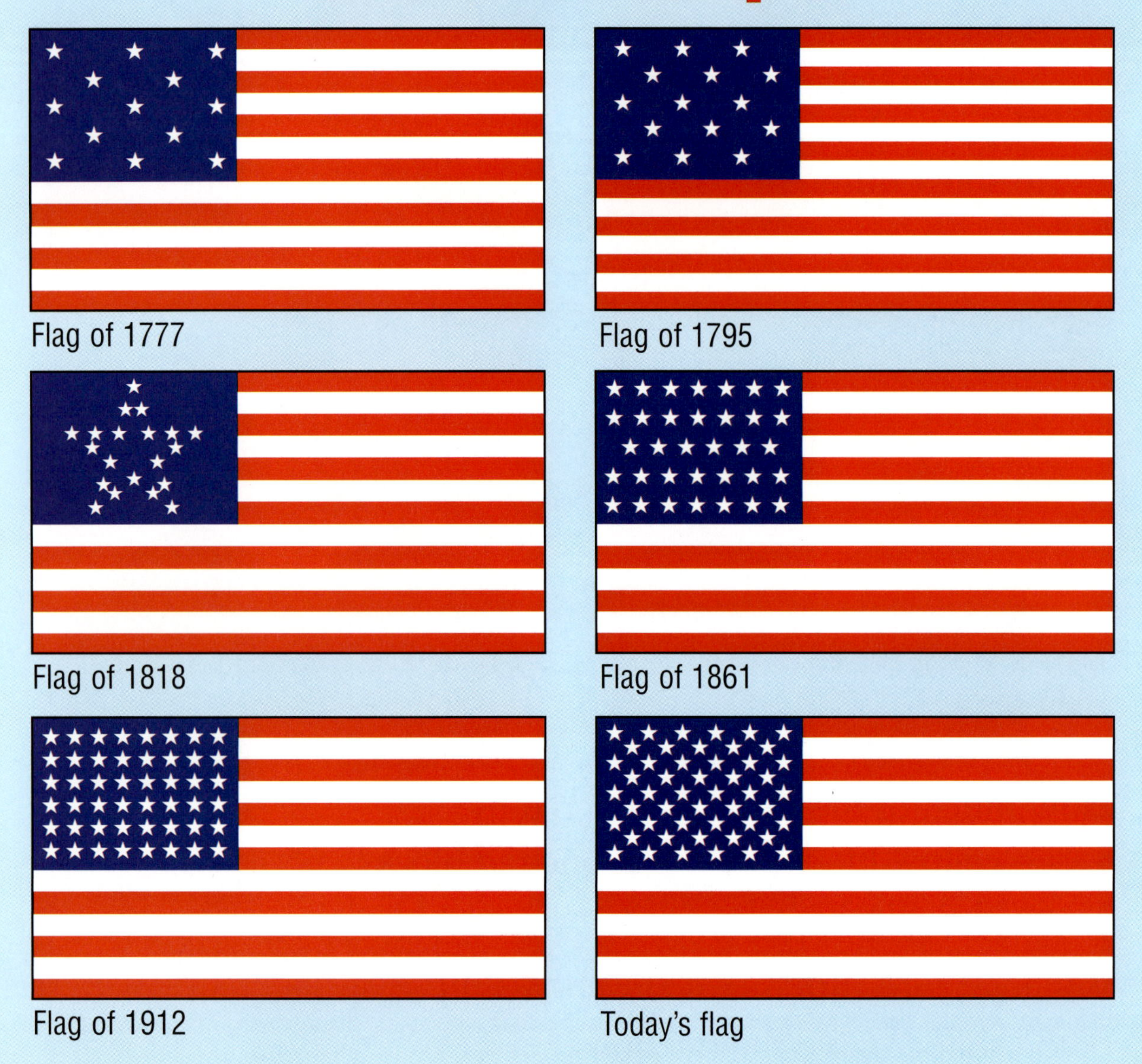

Flag of 1777

Flag of 1795

Flag of 1818

Flag of 1861

Flag of 1912

Today's flag

1. Describe how the stars are arranged on each flag.
2. Describe how you would find the total number of stars on each flag.
3. For which flags would it be easy to use multiplication to find the total? How do you know?
4. How would you use multiplication to find the number of stars on today's flag?
5. Design a flag that has 52 stars.

Myrna and Matt wrote the numbers from 1 to 50 on cards. Then they listed the factors of each number.

23	1, 23
24	1, 2, 3, 4, 6, 8, 12, 24
25	1, 5, 25
26	1, 2, 13, 26

1. How would you find all the factors of 27?
2. a. Which number between 1 and 50 do you think has the most factors? How would you check?
   b. Work in groups to find out and list the factors of all the numbers from 1 to 50.

Myrna and Matt decided to make a graph to show what they discovered.

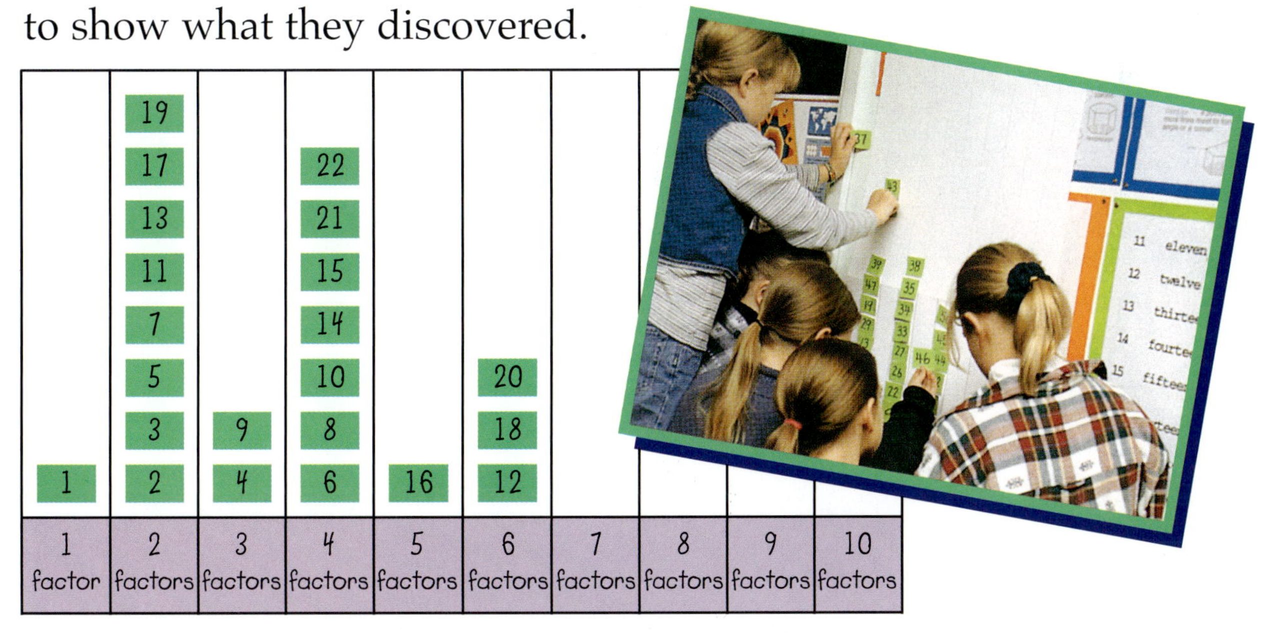

3. In which columns will Matt and Myrna put 23 24 25 26 ?
4. Work in groups to make a graph like Matt and Myrna's.
5. Which columns in your graph have the most factor cards? What other things did you discover?
6. What other questions can you investigate?

1. If you bought 6 loaves of bread, would you pay more or less than $4? Describe how you figured it out.
2. Suppose you had $5. Could you buy 6 quarts of milk? How do you know?
3. How could you figure out the cost of 6 tubs of margarine?
4. Which problems were you able to do easily in your head?
5. Look at each item. If you bought 5 of each item, which totals could you figure out easily in your head?
6. Which kinds of numbers do you find easy to multiply in your head? Explain why.

Some students decided to color grids to help them figure out the answer to 6 X 45.

1. Describe how Tanya colored her grid. How did she use it to figure out the answer?

2. How do you think Earl thought about the problem? Try to solve the problem Earl's way.

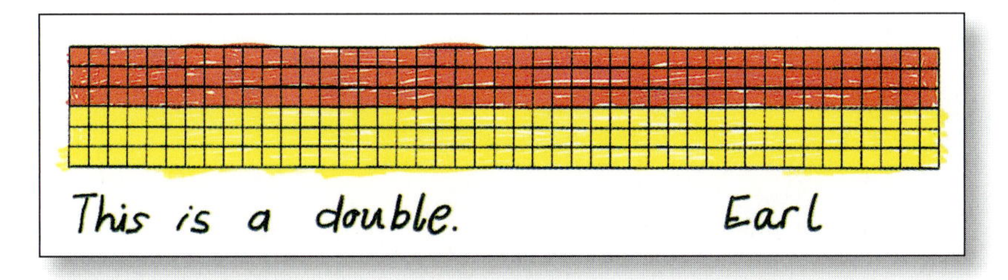

3. What are some different ways you could use Dana's picture to figure out the total?

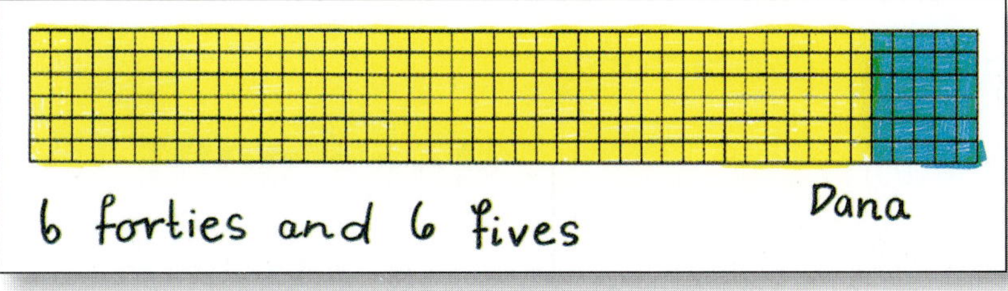

4. How could you find the cost of:

a. 4

b. 4

c. 5

The Dynacar factory produces cars 24 hours a day, 7 days a week.

1. Explain how you would figure out the number of cars built in

   a. one hour   b. one day   c. 100 days

2. How long do you think the Dynacar factory would take to make 1,000 cars?

3. Make up some other questions about the time it takes to produce cars at the Dynacar factory.

4. Make up some questions about these newspaper headlines. Challenge a partner to figure out the answers.

Donations pour in at the rate of $1,200 every day

Every week the population of Mechanicsville increases by 30 people

RACING PULSE

For this experiment, work with a partner and use a stopwatch.

Do all 4 activities.

After each activity below, ask your partner to feel your pulse and count the number of times your heart beats in 30 seconds.

1. Sit still for 5 minutes.
2. Walk 100 yards at your normal pace.
3. Run 100 yards as fast as you can.
4. Step onto a low bench 30 times.

The pulse rate tells the number of times a person's heart beats **per minute**. We feel the beats of the heart when we press against an artery in our wrist. This "pulse" can be felt in many parts of the body.

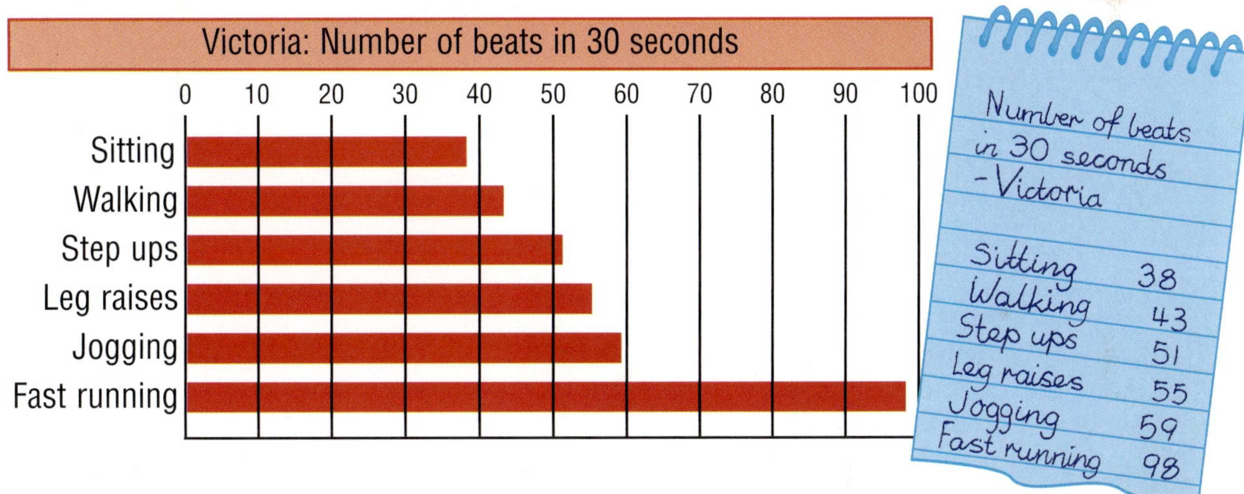

1. Use the graph to calculate Victoria's pulse rate after each activity. (Remember, the pulse rate is the number of beats in 60 seconds.)
2. Work with your partner to calculate what your pulse rate was after each activity.
   a. Which activity had the highest pulse rate?
   b. Which activity had the lowest pulse rate?
3. Make a graph of your own pulse count after 30 seconds.

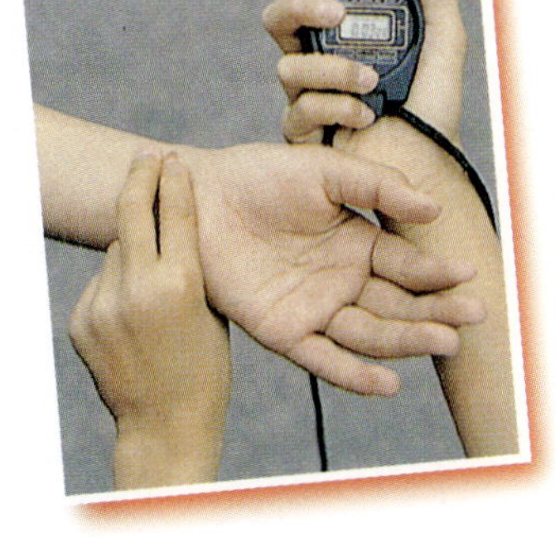

Washington ... New Carrollton ... Baltimore ... Wilmington ... Philadelphia ... Metropark ... Newark ... New York
WEEKDAY SERVICE

Train number ➡	10	12	14	16	18	20	26	32	34	38	48	50	52	54	56
Depart															
Washington DC	5:35 a.m.	6:00 a.m.	6:45 a.m.	7:00 a.m.	8:00 a.m.	9:00 a.m.	11:00 a.m.	12 noon	1:00 p.m.	3:00 p.m.	3:45 p.m.	4:00 p.m.	5:00 p.m.	6:00 p.m.	7:00 p.m.
New Carrollton, MD	5:44 a.m.	6:09 a.m.	6:54 a.m.	7:09 a.m.	8:09 a.m.	9:09 a.m.		12:09 p.m.				4:09 p.m.		6:09 p.m.	7:09 p.m.
BWI Airport Rail Sta., MD	5:59 a.m.	6:24 a.m.		7:24 a.m.		9:24 a.m.	11:22 a.m.		1:22 p.m.	3:22 p.m.			5:22 p.m.		7:24 p.m.
Baltimore, MD	6:12 a.m.	6:37 a.m.		7:37 a.m.	8:33 a.m.	9:37 a.m.	11:34 a.m.	12:33 p.m.	1:35 p.m.	3:35 p.m.		4:33 p.m.	5:35 p.m.	6:33 p.m.	7:37 p.m.
Wilmington, DE	6:56 a.m.	7:22 a.m.		8:22 a.m.	9:18 a.m.	10:22 a.m.	12:19 p.m.	1:18 p.m.	2:20 p.m.	4:20 p.m.		5:18 p.m.	6:20 p.m.	7:18 p.m.	8:22 p.m.
Philadelphia, PA	7:19 a.m.	7:45 a.m.	8:21 a.m.	8:45 a.m.	9:40 a.m.	10:45 a.m.	12:41 p.m.	1:40 p.m.	2:42 p.m.	4:42 p.m.	5:20 p.m.	5:40 p.m.	6:42 p.m.	7:40 p.m.	8:43 p.m.
Trenton, NJ											5:47 p.m.				
Princeton Jct., NJ											5:55 p.m.				
Metropark, NJ					10:28 a.m.		1:30 p.m.		3:30 p.m.	5:30 p.m.	6:11 p.m.	6:28 p.m.		8:28 p.m.	9:31 p.m.
Newark, NJ	8:15 a.m.	8:41 a.m.		9:41 a.m.	10:41 a.m.	11:41 a.m.	1:43 p.m.	2:39 p.m.	3:43 p.m.	5:43 p.m.	6:24 p.m.	6:41 p.m.	7:39 p.m.	8:41 p.m.	9:44 p.m.
New York, NY	8:35 a.m.	8:59 a.m.	9:29 a.m.	9:59 a.m.	10:59 a.m.	11:59 a.m.	1:59 p.m.	2:59 p.m.	3:59 p.m.	5:59 p.m.	6:40 p.m.	6:59 p.m.	7:59 p.m.	8:59 p.m.	9:59 p.m.

1. What are the departure times of trains leaving Baltimore
   a. before 8:00 a.m.?
   b. between 4:30 p.m. and 6:30 p.m.?
2. What is the usual length of time it takes to travel from Washington to New York? How did you decide?
3. Which train takes the shortest length of time to travel from Washington to New York? How do you know?
4. Plan a trip from Washington to New York so you can stop in Philadelphia for lunch. Write an **itinerary** for your trip.
5. If you live in Philadelphia, what time do you need to catch the train to
   a. be in New York by noon?
   b. travel to Princeton, N.J.?

BUILDING A BOOKCASE

1. What is the total length of the lumber needed to build the frame of the bookcase? Explain how you know.
2. How would you buy the lumber? (You can cut it to make the frame.) The lumber is only sold in pieces that are 6′, 8′, 10′, or 12′ long.
3. Five shelves are planned for the bookcase. How much lumber do you need for the entire bookcase?
4. What is the most economical way to buy the lumber for the bookcase?
5. Compare the prices of the 6′ and 12′ lengths of lumber. What is surprising about them?

Lumber prices	
6-foot lengths	$7.20
8-foot lengths	$9.60
10-foot lengths	$12.00
12-foot lengths	$15.00

You will need a rectangle of grid paper.

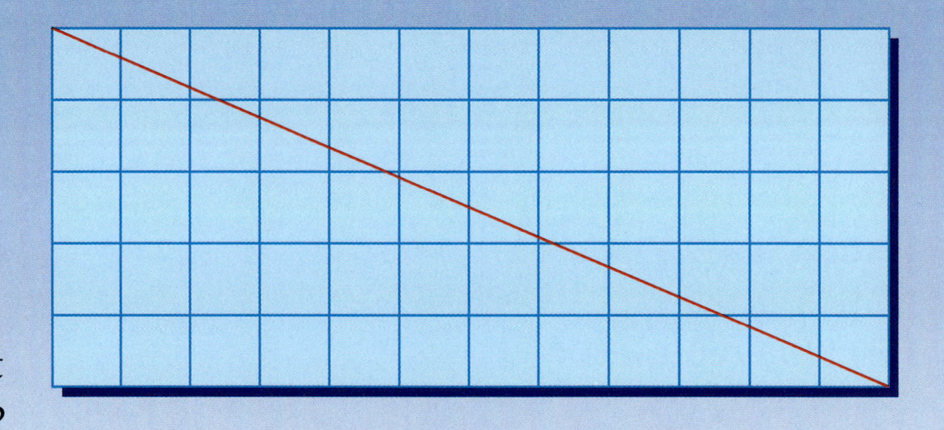

Jaime cut this rectangle into 2 triangles. What steps did he use?

1. What are the names of the shapes Jaime made when he rearranged his two triangles? (Sometimes he had to turn over a triangle.)

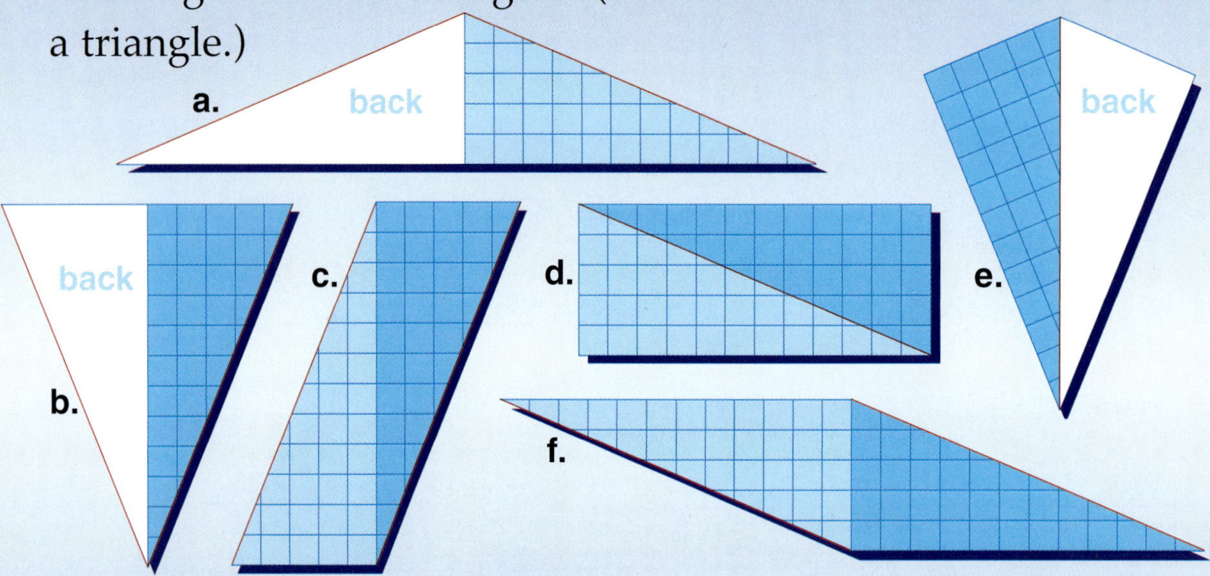

2. What is the same and what is different about Jaime's shapes?
3. Which shapes have the same perimeter as the rectangle Jaime started with?
4. Which shapes have a perimeter that is greater than the rectangle's perimeter?
5. Work with a partner and use a square of grid paper. Cut the square along one diagonal and make new shapes with the 2 triangles you get. Find out which shapes have the same perimeter.

There are several ways you could tie a ribbon around a large box.

1. Look at the dimensions of the box that the girl is holding. List its length, height, and width.
2. How much ribbon is required for Style A? (Allow an extra 24 inches for the bow.) How did you figure it out?

Style A

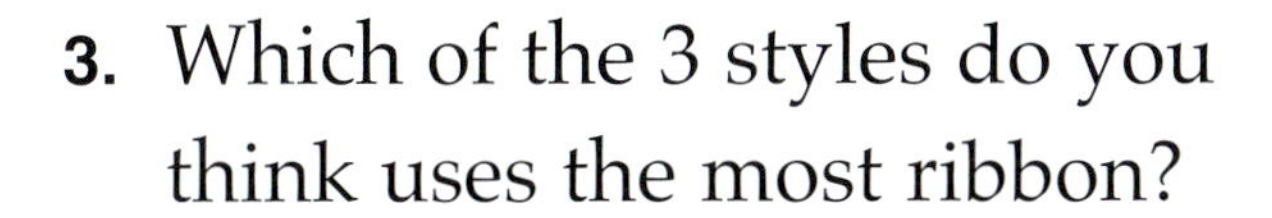

3. Which of the 3 styles do you think uses the most ribbon?
4. Write the length of ribbon needed for Styles B and C. (Don't allow for the bow.)
5. List the calculations you made when you figured out the ribbon length for Style C.
6. How would you find the distance around the middle of this cube-shaped box?

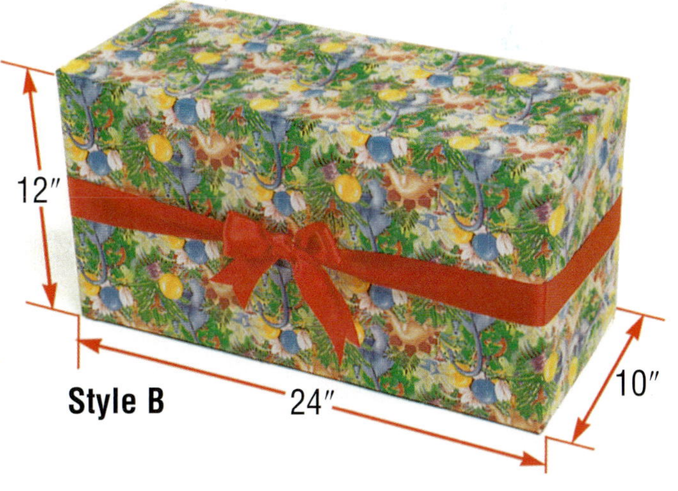

Style B

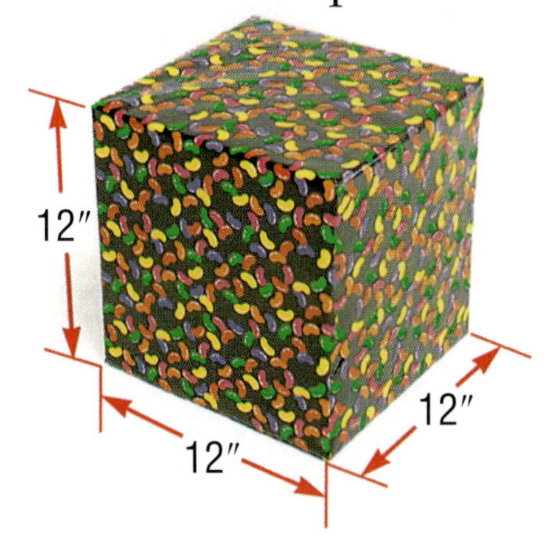

Style C

On Saturday, Jamal and Antonio split a pizza.

1. What fraction of the pizza did each boy eat if
   a. Jamal ate 2 pieces?
   b. Antonio ate 3 pieces?
2. On Saturday night, they each ate one more piece of pizza. What fraction of the whole pizza did each boy eat on Saturday? How did you decide?
3. What fraction of the pizza was left?
4. Jon, Jenny, Jeff, and Jolene shared 6 waffles.

   a. How many waffles did each person get?
   b. Jenny ate $1\frac{1}{4}$ waffles. She gave the rest of her share to Jon. How much did she give Jon? How many waffles did Jon eat? How did you decide?
5. If they made one more waffle to share, how many waffles would each person have eaten in total? Explain your thinking.

Sara and Matt were both given some money.
They drew pie graphs to show what happened to the money.

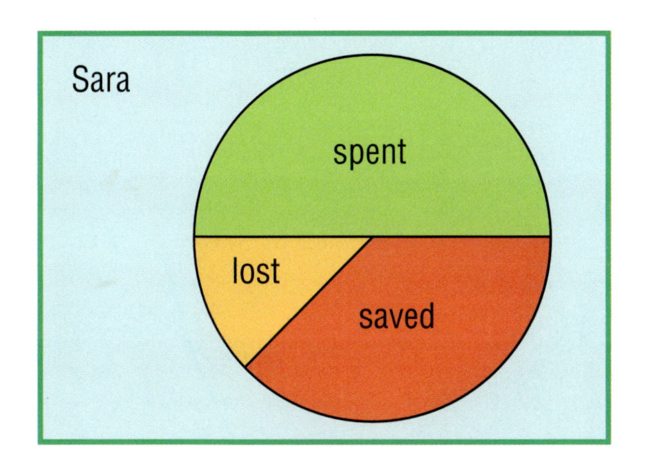

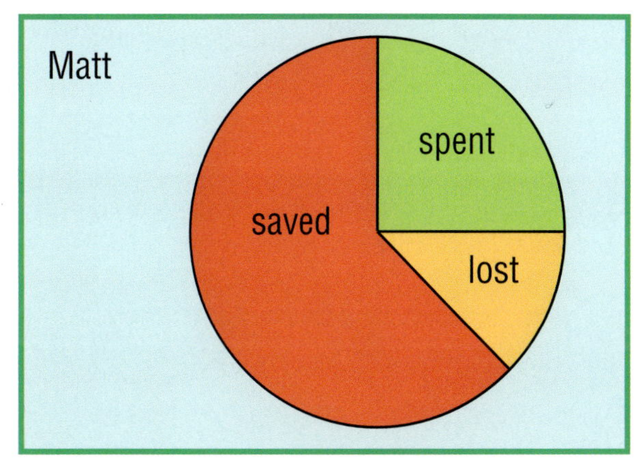

1. Could Matt have spent more than Sara?
   Explain your answer.
2. If Sara had been given \$8, about how much
   a. did she spend? **b.** did she lose?
3. If Matt had been given \$12, about how much
   a. did he spend? **b.** did he lose?

Daniel and Becca showed what they did with their money.

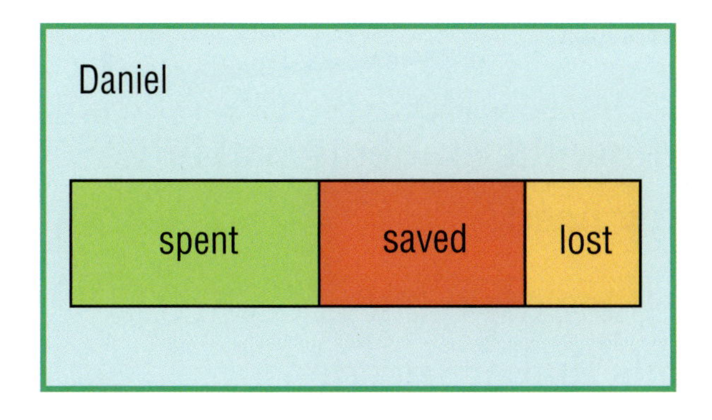

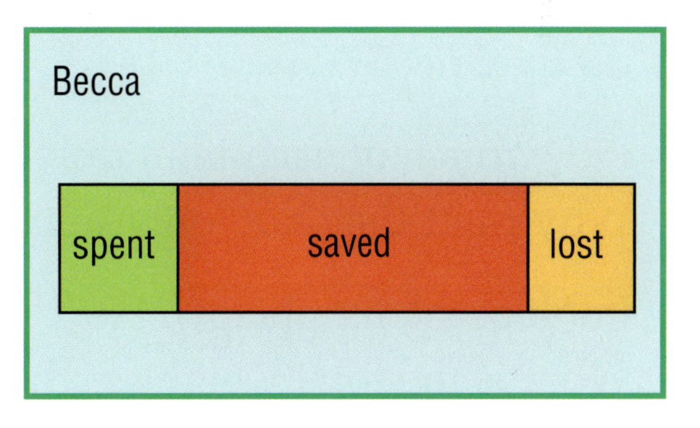

4. Who spent more, Daniel or Becca? Could it be either one?
   Explain your thinking.
5. Maria spent $\frac{1}{3}$ of her money. Tony spent $\frac{1}{4}$ of his money.
   Tony said he spent more than Maria. Could he be right?

Materials we need

Tagboard

$1\frac{1}{2}$ sheets for the front
$1\frac{1}{4}$ sheets for the back
$\frac{3}{4}$ sheet for the top
$\frac{3}{4}$ sheet for the left side
$1\frac{1}{2}$ sheets for the right side
$\frac{1}{2}$ sheet for lettering

Fancy Trim

$6\frac{2}{3}$ feet for the outside edges
$5\frac{2}{3}$ feet for around the opening

Paint

$4\frac{1}{2}$ ounces of red for the outside
$3\frac{1}{4}$ ounces of green for the inside
$1\frac{1}{2}$ ounces of yellow for the lettering

Some students decided to construct a puppet theater. They decided to make the theater out of **tagboard** and **fancy trim** and then paint it **red**.

The students made a list.

1. How much tagboard, fancy trim, and red paint do they need?

2. How much of each item do you think the students will buy? Why can't they buy only what they need?

3. Look at the lists below. How much brown paint is needed? What is the total weight of the nails that are needed? Explain your thinking.

a.

Brown Paint for Tree House
$1\frac{1}{2}$ quarts for the walls
$\frac{1}{2}$ quart for floor
$\frac{1}{2}$ quart for roof

b.

Nails for Tree House
$\frac{1}{4}$ pound 1" nails
$\frac{1}{2}$ pound $1\frac{1}{2}$" nails
$1\frac{1}{2}$ pounds 2" nails

Ayana and Rita have been adding and subtracting fractions.

Ayana

$\frac{3}{8} + \frac{1}{2} = \frac{4}{8}$

$\frac{1}{4} + \frac{1}{2} = \frac{2}{4}$

$\frac{3}{6} - \frac{1}{3} = \frac{2}{6}$

$\frac{7}{8} - \frac{3}{4} = \frac{4}{8}$

Rita

$\frac{1}{2} + \frac{1}{4} = \frac{2}{6}$

$\frac{1}{4} + \frac{3}{4} = \frac{4}{8}$

$\frac{3}{4} - \frac{1}{2} = \frac{2}{2}$

$\frac{5}{8} - \frac{1}{4} = \frac{4}{4}$

1. Look at the answers Ayana and Rita wrote.
   a. Are their answers reasonable?
   b. Explain your thinking. You can draw pictures, use words, or use the fraction strips to help.

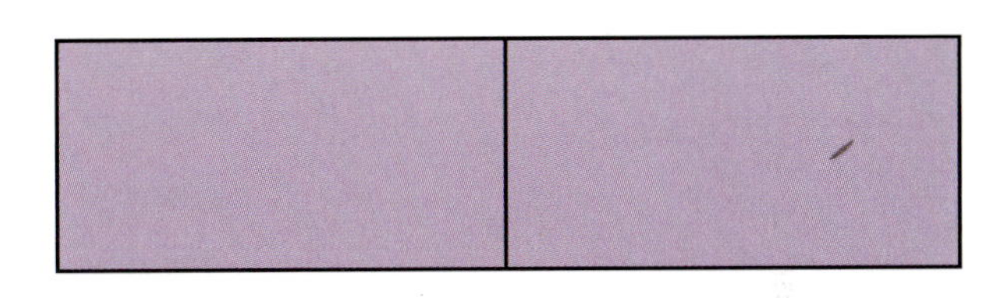

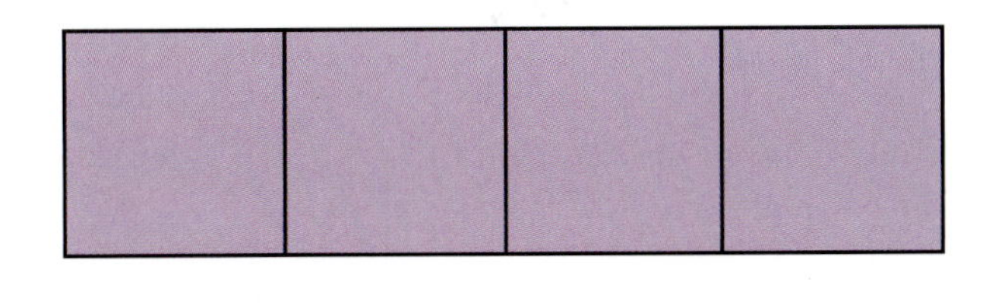

2. Use the fraction strips, or your own method, to find the answers for Ayana and Rita. Explain how you did it.

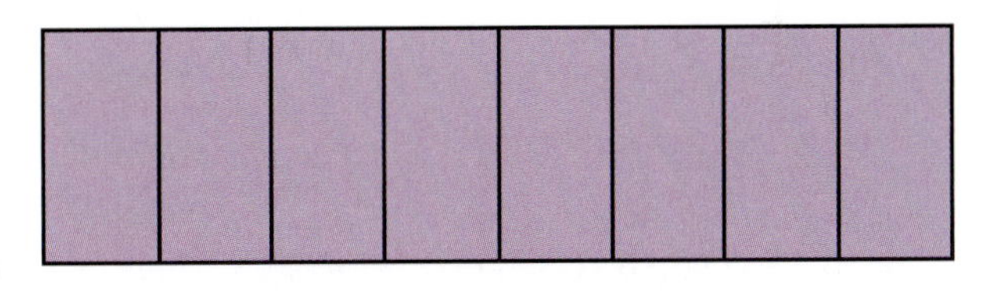

3. Make up 4 of your own addition and subtraction fraction questions. Use the strips to help.

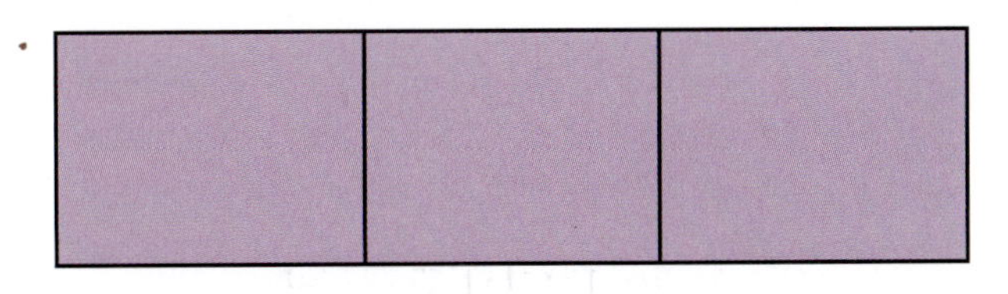

4. Find the answers to your questions, then challenge a partner to figure them out.

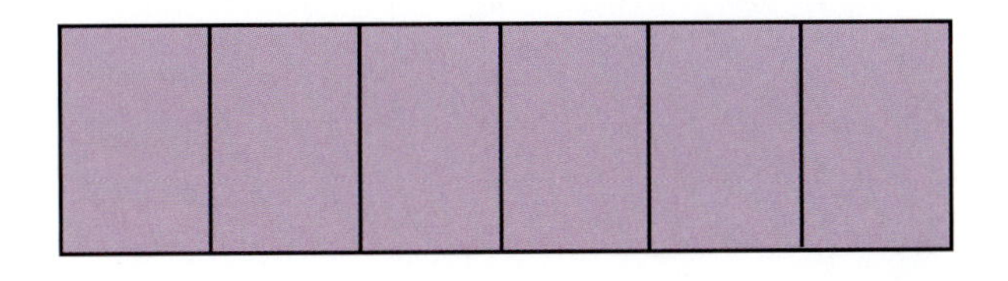

1. Mrs. Smith, Mr. Chau, and Mrs. Quinn went to the museum together. They paid admission for a group of three and then split the cost. How much did each person pay?

2. Figure out the price per person for a group of
   a. 2 adults **b.** 4 adults

3. Explain how you use multiplication and division to find the price per person for each group admission price.

4. How does the price per person change as groups get bigger?

5. Look at the Raging Reptiles admission prices. What pattern do you see? What might the prices be for
   a. 4 adults? **b.** 5 adults?
   Explain your thinking.

MUSEUM ADMISSION PRICES

1 adult $12

GROUP

2 adults $20

3 adults $27

4 adults $34

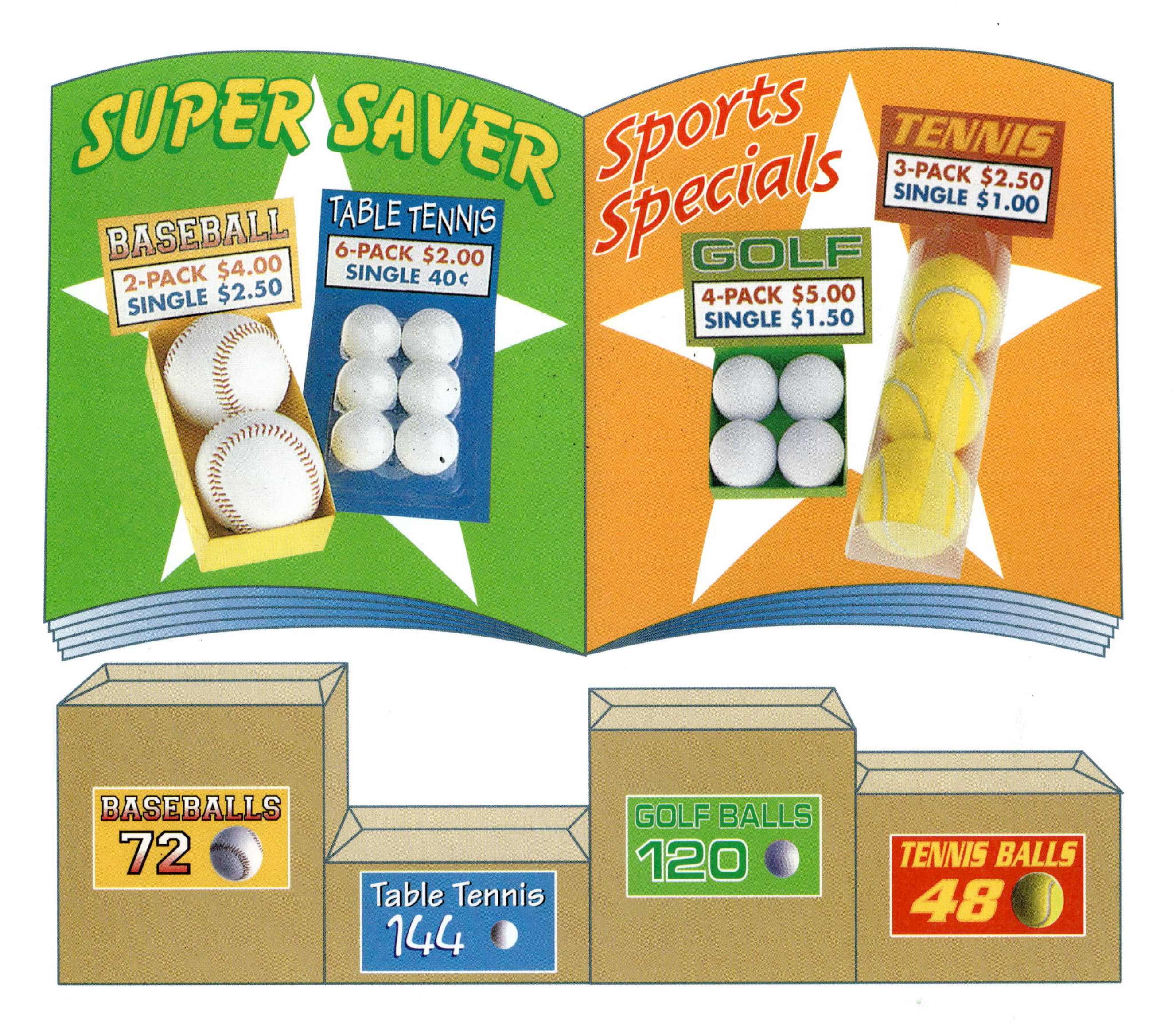

1. How many 4-packs of golf balls could you fill from the large box of golf balls? How did you figure out the answer?
2. In one month, 38 packages of table tennis balls were sold. Is that more or fewer table tennis balls than the large box holds? How do you know?
3. Use the picture to make up more problems about the balls.
4. Write the problems on the board.
   - **a.** Which are multiplication problems?
   - **b.** Which are division problems?
5. Solve the problems as a class.

Members of the school band formed groups to raise money.

1. Which student raised the most money?
   Which group was the student in?
2. Who raised the least amount?
   Which group was that student in?
3. How much did each group raise?
   Which group raised the most?
   Were you surprised by this?
4. Look at the Buglers' board. How many students are in that group? If the total amount they raised is $72, how can you figure out how much they raised per person?
5. Find out how much the Drummers raised per person.
6. If they decided to share the money equally, how much would each group get?
7. If the Drummers and the Buglers combined the money they raised, how much would they have raised per person?

1. How high would Airplane A be if it climbed
   a. 1 foot? **b.** 10 feet?
   c. 100 feet? **d.** 1,000 feet?

2. How high would Airplane B be if it descended
   a. 1 foot? **b.** 10 feet?
   c. 100 feet? **d.** 1,000 feet?

3. If Airplane C climbed 100 feet every 10 seconds, what would its altitude be after
   a. 10 seconds? **b.** 20 seconds?
   c. 40 seconds? **d.** 1 minute?

4. If Airplane D descended 1,000 feet every 30 seconds, what would its altitude be after
   a. 30 seconds? **b.** 1 minute?
   c. 2 minutes? **d.** 4 minutes?

1. Read what you see on the number expander in Step 1. What pattern can you see in the words?

Step 1

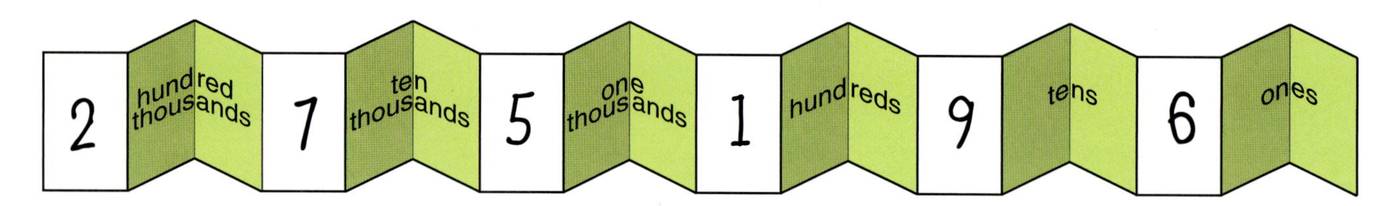

2. Read what you see on the expander in Step 2. Why do you think part of the expander is closed?

Step 2

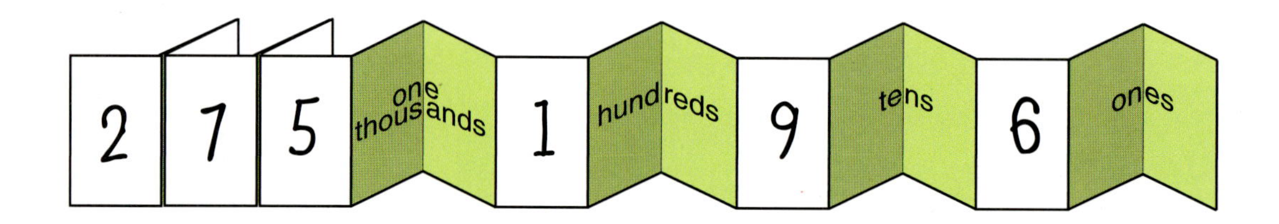

3. Read what you see in Step 3. What number is the expander showing?

Step 3

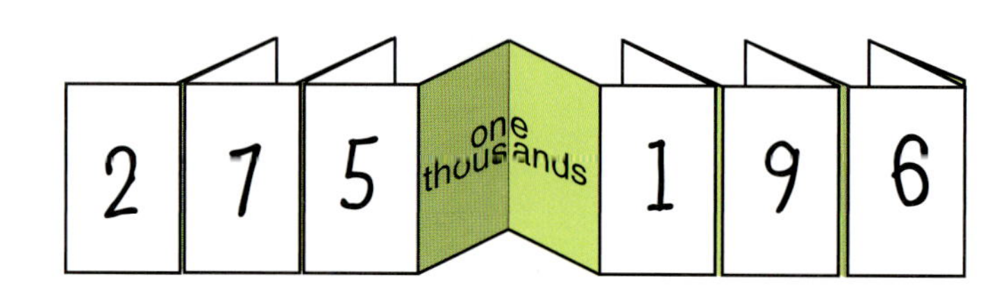

4. Look at the expanders below. Read each number aloud. What do you notice when you say the numbers?

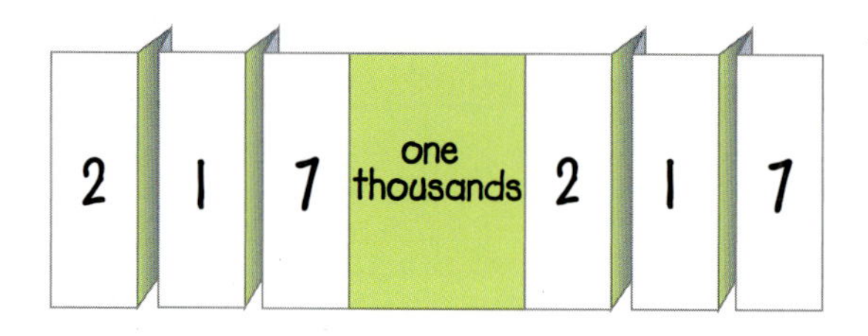

1. Which two states have the largest road networks?
2. Estimate the difference between the two largest networks.
3. Which states have a road network greater than 120,000 miles?
4. List the states in order from the largest to the smallest road network.
5. Which two states have the least difference between the length of their networks? Estimate the difference.
6. Why do you think some states with a large area are not on the list?

THE TOP TEN ROAD NETWORKS

State	Length in miles	State	Length in miles
California	169,201	Minnesota	129,959
Florida	112,808	Missouri	121,787
Illinois	136,965	Ohio	113,823
Kansas	133,256	Pennsylvania	117,038
Michigan	117,659	Texas	294,142

Source: The Top Ten of Everything 1996

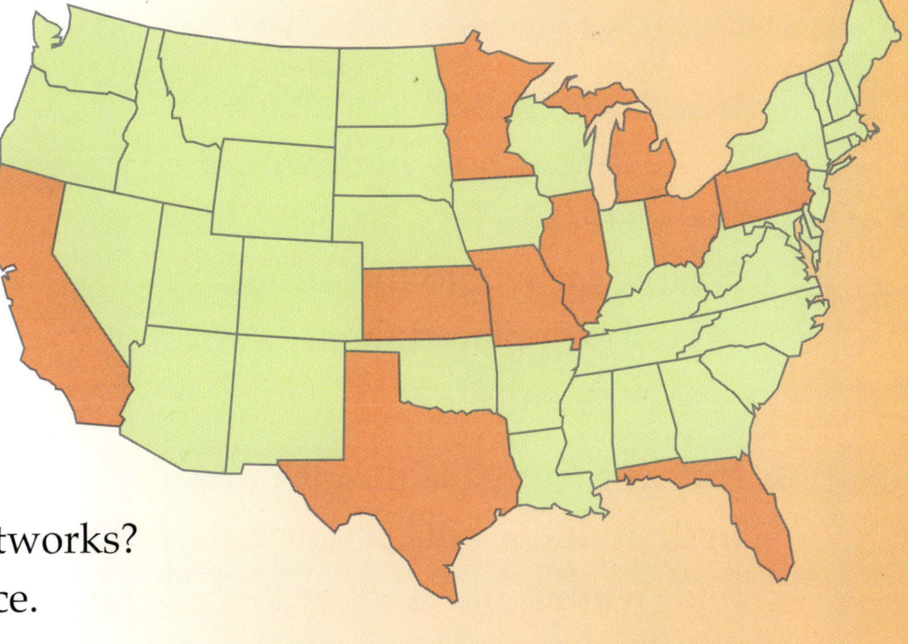

Making a Million

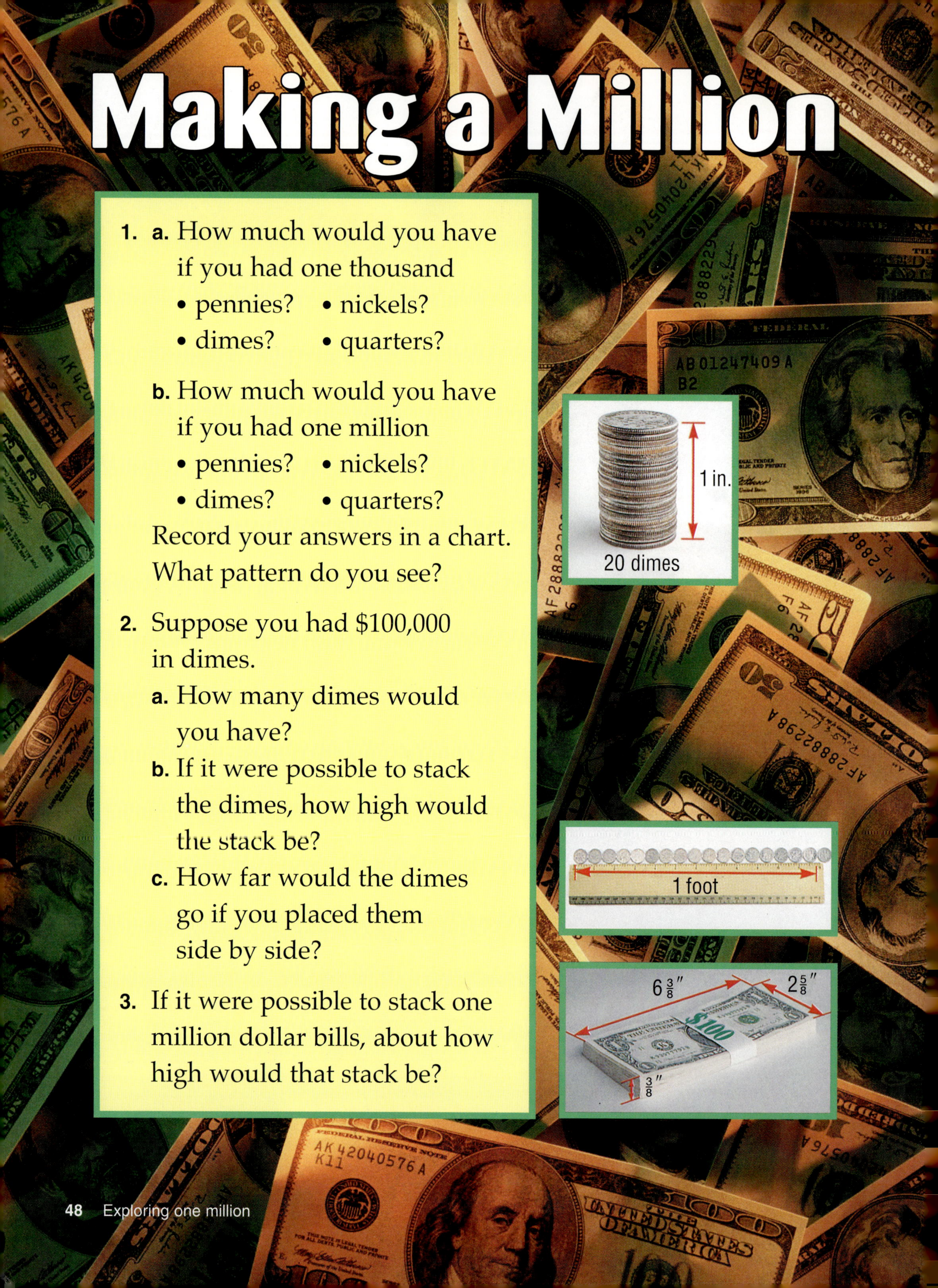

1. a. How much would you have if you had one thousand
   - pennies?
   - nickels?
   - dimes?
   - quarters?

   b. How much would you have if you had one million
   - pennies?
   - nickels?
   - dimes?
   - quarters?

   Record your answers in a chart. What pattern do you see?

2. Suppose you had $100,000 in dimes.

   a. How many dimes would you have?

   b. If it were possible to stack the dimes, how high would the stack be?

   c. How far would the dimes go if you placed them side by side?

3. If it were possible to stack one million dollar bills, about how high would that stack be?

MRS. BARKER'S BAKERY

SINGLES	45¢
TWIN PACK	85¢
FOUR SQUARE	$1.60
HALF DOZEN	$2.20

1. Suppose you had $5.
   a. How many single muffins do you think you could buy? How do you know?
   b. How many twin packs do you think you could buy? How would you figure it out?
   c. What is the greatest number of muffins you could buy? Explain your thinking.

2. Suppose Mrs. Barker had 20 muffins left. How much money would she make if she sold the twenty muffins:
   a. as singles? b. in twin packs? c. in four squares?

3. Mrs. Barker hoped to sell 60 muffins. How much money would she make if she sold them as singles rather than in half-dozen packs?

1. Kim wants to buy the Electronic Easel. How can she do this and use up most of her coins?

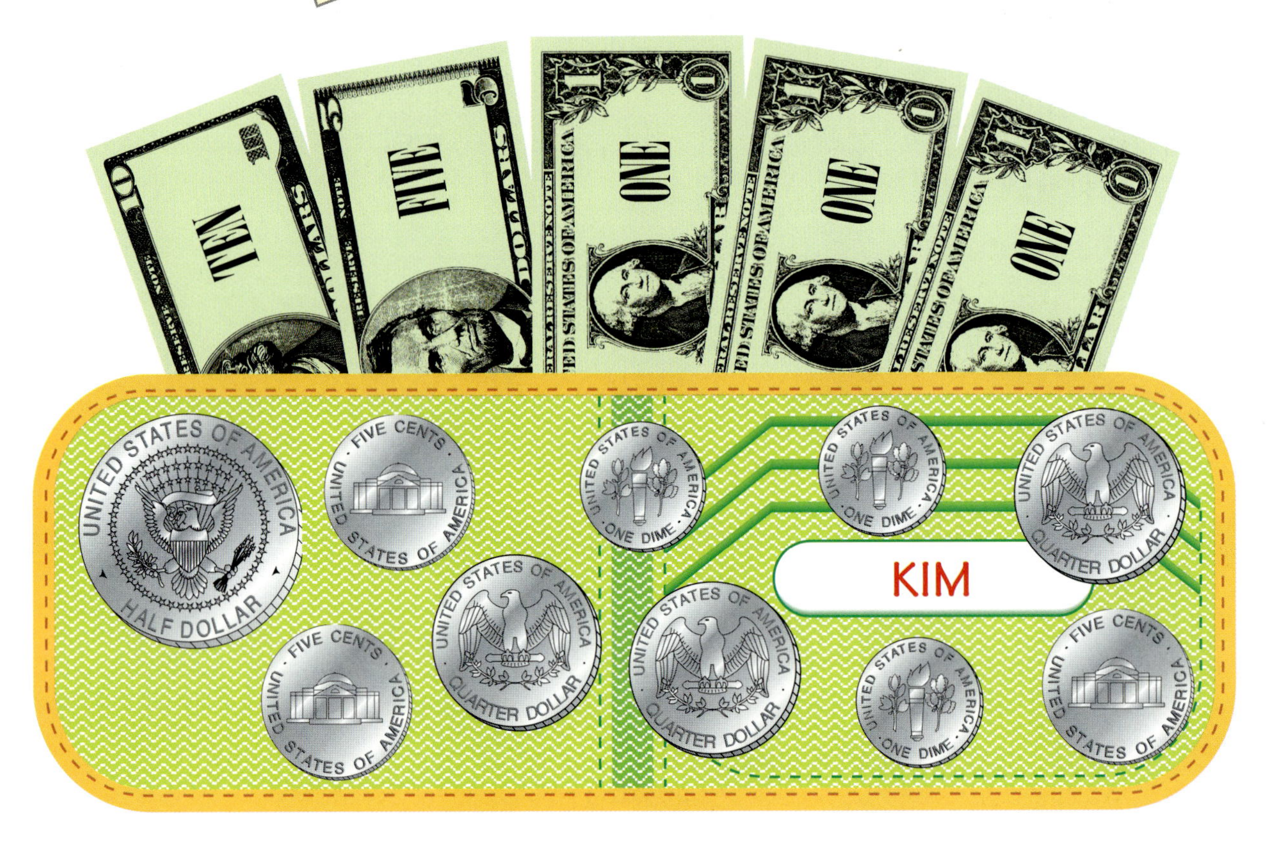

2. How much money will Kim have left?

3. Craig also wants to buy an Electronic Easel. Suppose he hands the salesperson a $20 bill. What are some possible combinations for his change?

4. Chuck has 3 five-dollar bills, 4 one-dollar bills, 6 quarters, 4 dimes, and 3 nickels. How could he buy the Electronic Easel and use up most of his coins?

David, Rocio, and Darrel made graphs to show how they spent $100.

1. Who spent
   a. $15 on carry-out food?
   b. between $20 and $25 on movies?
   c. $15 on gifts?
   d. about $\frac{1}{4}$ of the $100 on carry-out food?
2. Who spent the most on
   a. travel?
   b. gifts?
   c. carry-out food?
3. Did any person spend more than $\frac{1}{3}$ of the $100 on one kind of item? How do you know?
4. Which graph is easiest to read if you want to know
   a. exact amounts?
   b. fractions of the total?

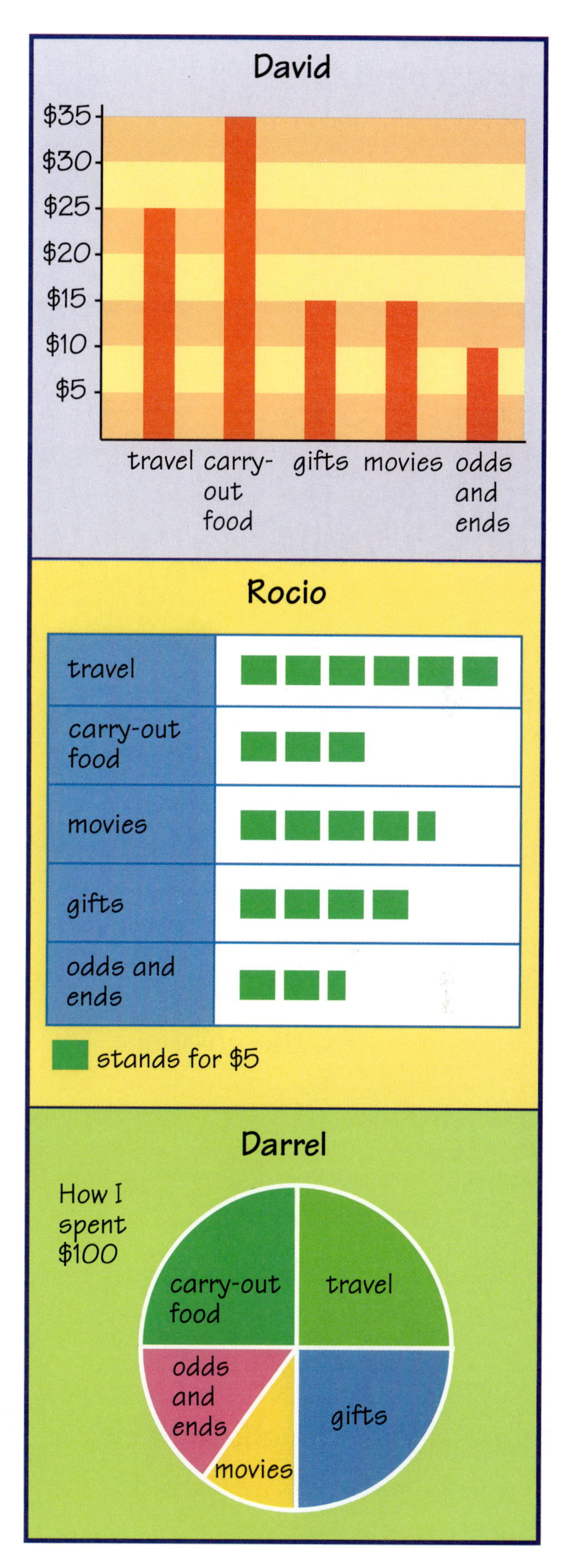

1. Describe 3 different ways that you could buy 6 quarts of paint. What is the cheapest way? How did you figure it out?

2. If you need 5 quarts of paint, what is
   a. the cheapest way to buy it?
   b. the most expensive way to buy it?
   c. the difference between the cheapest and the most expensive way? Explain how you found the answers.

3. If you have $60, what is the most paint you can get for your money?

4. Painting a large room will require $2\frac{1}{2}$ gallons of blue paint, 6 quarts of green paint, and 1 pint of yellow paint. Explain how you would find the total cost.

1. What is shown on the four charts? What is the same and what is different?

2. Work in groups.
   a. Make up a set of money problems to match the four pictures. Make each unit one dollar.
   b. Share the sets of problems as a class.
   c. What do you notice about the answers to each set of problems?

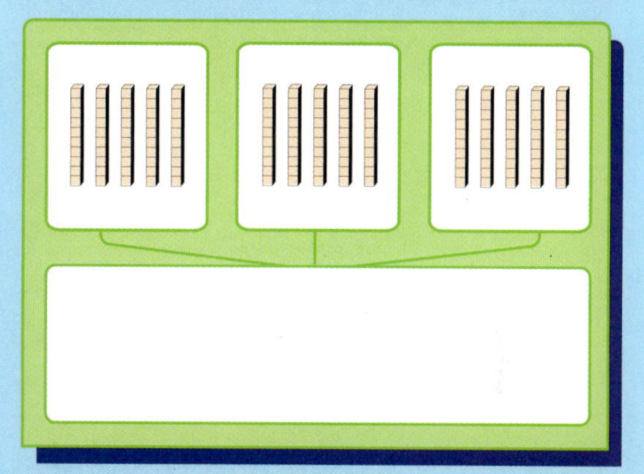

3. Now make up a set of length problems to match the four pictures. Make each unit one foot.

4. Arrange each set of problems in order. Write about the pattern you see.

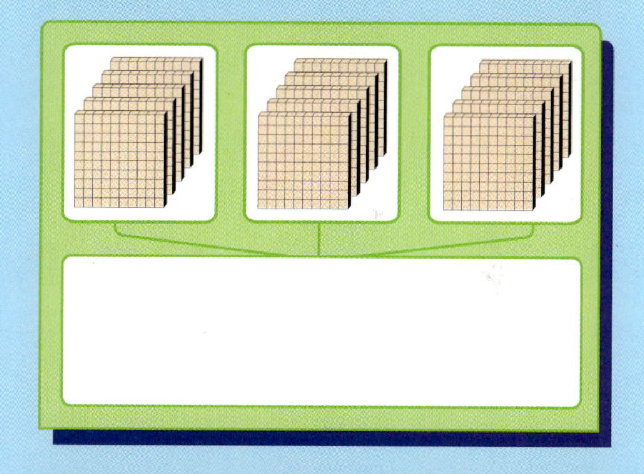

5. Copy and complete these number sentences. Write about the patterns you see.

$3 \times 6 =$ _____	$8 \times 7 =$ _____
$3 \times 60 =$ _____	$8 \times 70 =$ _____
$3 \times 600 =$ _____	$8 \times 700 =$ _____
$3 \times 6{,}000 =$ _____	$8 \times 7{,}000 =$ _____

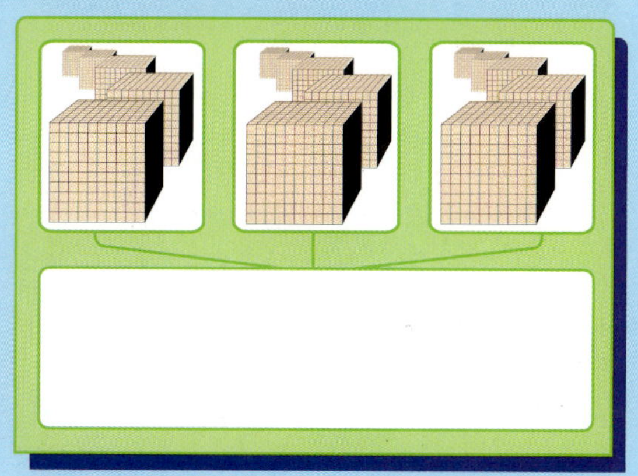

1. What would you pay for each of these purchases?
   a. 10 T-shirts b. 15 T-shirts c. 18 T-shirts
   d. 10 sweatshirts e. 15 sweatshirts f. 18 sweatshirts
   Explain how you found the answers.

2. What would you pay for these purchases?
   a. 30 sweatshirts
   b. 25 sweatshirts
   c. a sweatshirt for each person in your class

3. Make a table like the one below. Go up to 50. Write the amount you will pay for each purchase. What patterns do you see in your chart?

Number bought	$3 T-shirts	$30 Sweatshirts
40		
41		
42		

Ms. Duncan made this "time saver" to help her calculate the number of bottles in loads of four-packs.

NUMBER OF FOUR-PACKS		Ones									
			1	2	3	4	5	6	7	8	9
Tens		0	4	8	12	16	20	24	28	32	36
	10	40	44	48	52	56	60	64	68	72	76
	20	80	84	88	92	96	100	104	108	112	116
	30	120	124	128	132	136	140	144	148	152	156
	40	160	164	168	172	176	180	184	188	192	196

1. How many bottles are there in
   a. 2 four-packs? **b.** 4 four-packs? **c.** 6 four-packs?
   d. 20 four-packs? **e.** 40 four-packs? **f.** 60 four-packs?
   How could you use the "time saver" to find the answers?

2. How many bottles are there in
   a. 24 four-packs? **b.** 46 four-packs?
   How did you find the answers?

3. How can you use the "time saver" to find the total number of bottles in 27 four-packs?

4. Find the number of bottles in
   a. 32 four-packs **b.** 41 four-packs **c.** 49 four-packs
   Explain how you used the "time saver" to find the answers.

5. How do you think Ms. Duncan made up her "time saver"?

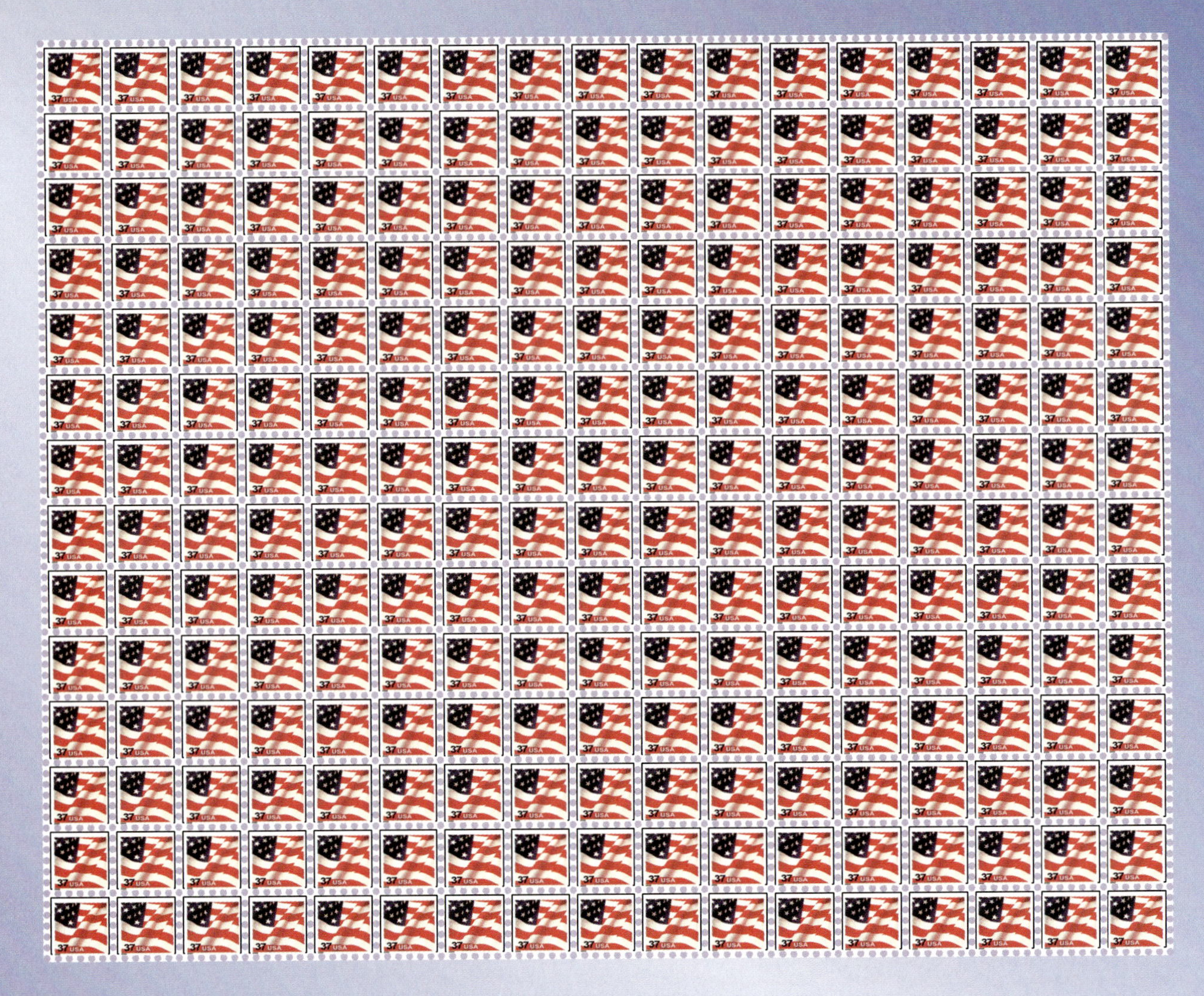

1. How would you find the total number of stamps without counting?
2. Look at the picture. Think of two 2-digit numbers for which you know the product. Cover up an array shown by those two numbers. How many stamps did you cover?
3. Select another array where you know the product of the two 2-digit numbers. Cover up that array and explain how you know the number of stamps that are covered.
4. Write the two numbers that describe the whole array. What is the total number of stamps shown in the array? Explain how you know.

Jerrard is covering his table top with 1-inch square tiles.

1. What is the shape of the table? What are the dimensions of the table? What words can you use to describe the dimensions?

2. Estimate the number of 1-inch square tiles needed to cover the table top. How did you make your estimate?

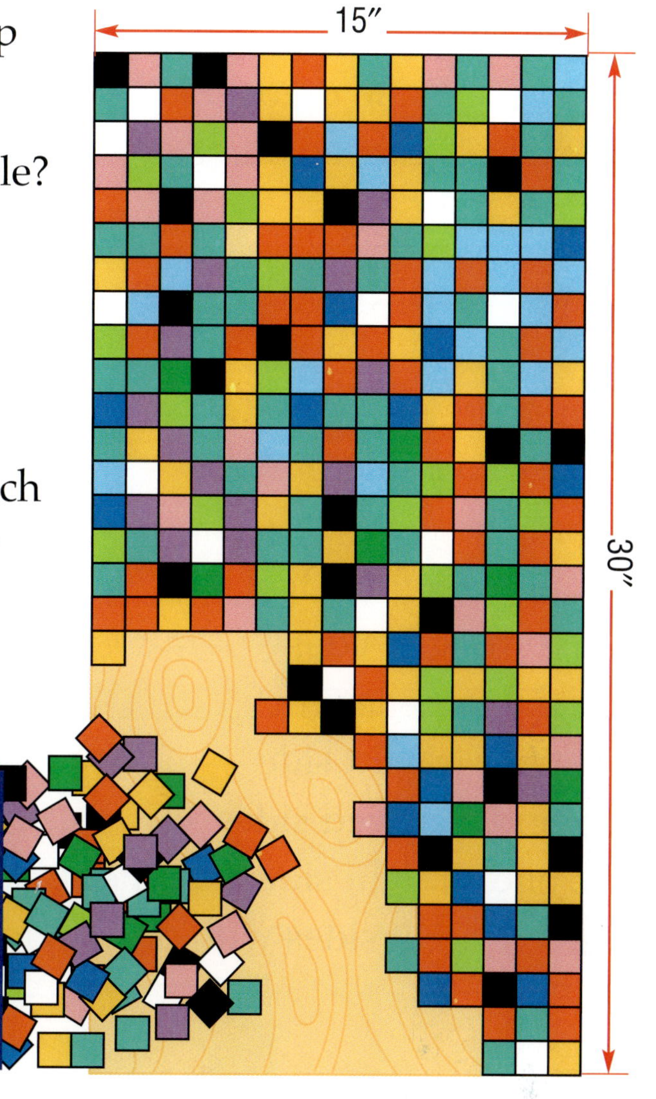

To find the **area of a rectangle**, multiply the length of the base by the height of the rectangle. Area is given in square units, such as square inches, square feet, or square yards.

3. Give the area of the table top in square inches.

4. Calculate the area of each of these shapes.

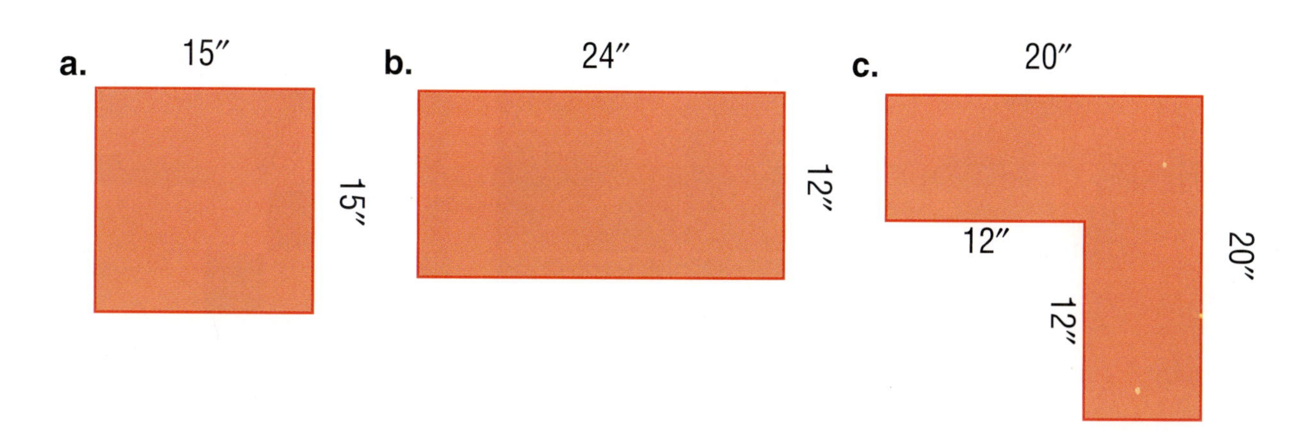

1. Find shapes in the design that are made up of smaller shapes. Describe them.

2. Find the 5 sets of shapes in the design. How many shapes are there in each set? Point to a shape that has

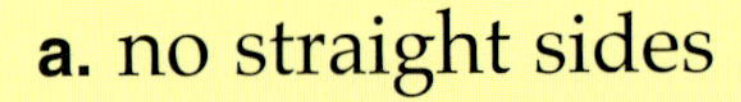

a. no straight sides
b. 3 straight sides the same length
c. one straight side and 2 curved sides

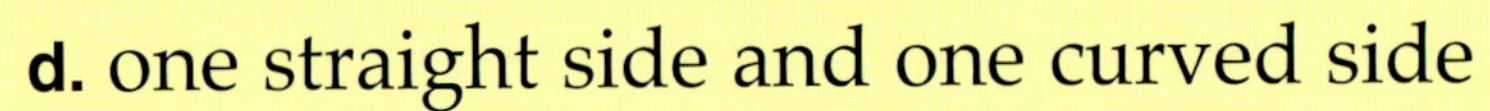

d. one straight side and one curved side

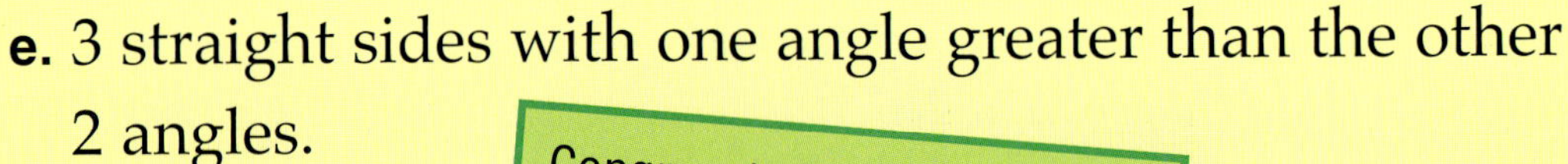

e. 3 straight sides with one angle greater than the other 2 angles.

Congruent shapes are exactly the same size and shape.

3. Look at each quilt design. Describe the shapes within each design that are congruent.

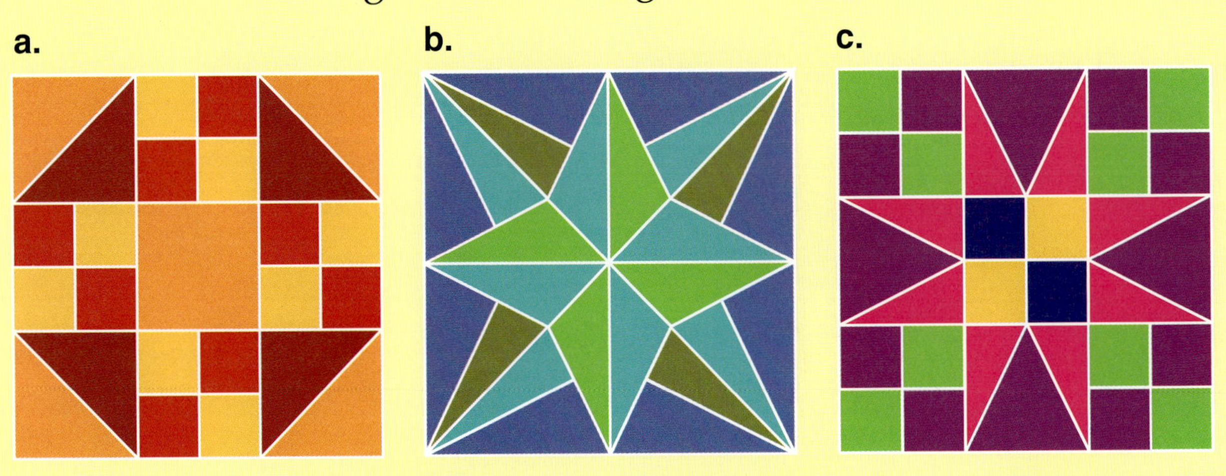

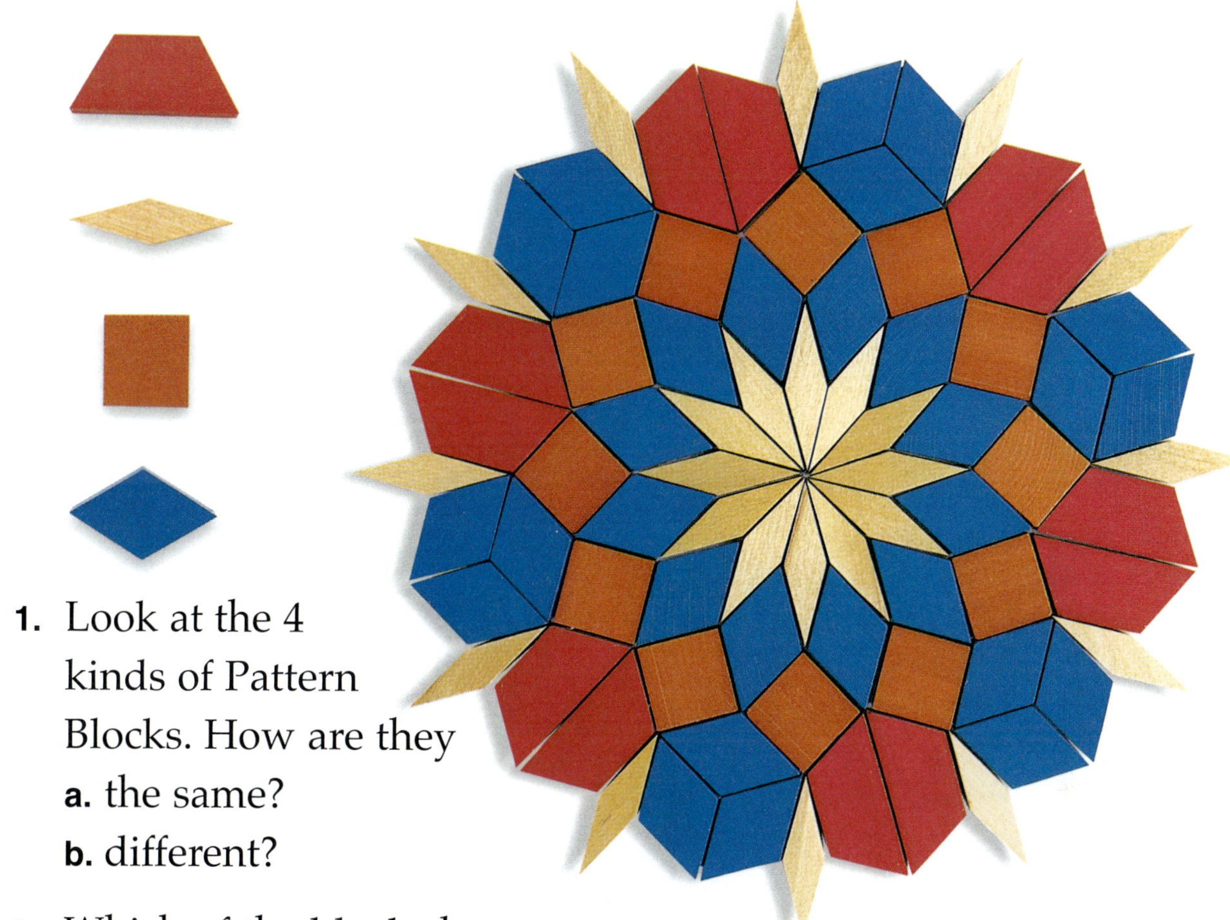

1. Look at the 4 kinds of Pattern Blocks. How are they
   a. the same?
   b. different?

2. Which of the blocks have
   a. both pairs of opposite sides *parallel*?
   b. *two* angles that are less than a right angle?

3. Matt made a paper cutout of a blue Pattern Block.
   a. What did he do in Step 1?
   b. In Step 2, Matt rearranged the 4 torn corners. What do you notice about the corners now?

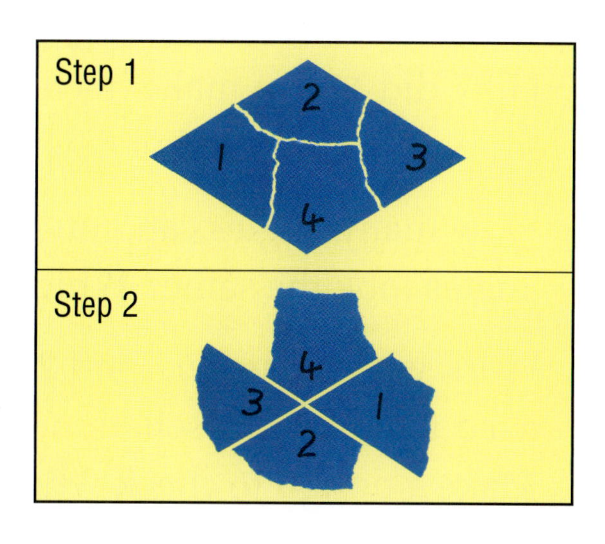

4. Make copies of the 4 Pattern Blocks. Tear off the corners of each paper copy. Rearrange them as in Step 2. What do you notice?

Milly put 2 Pattern Blocks together to make a new shape. She used her new shape to make a design where all the shapes fit together with no gaps.

A design where all the parts fit together with no gaps is called a tessellation.

1. Describe how Milly made this tessellation.

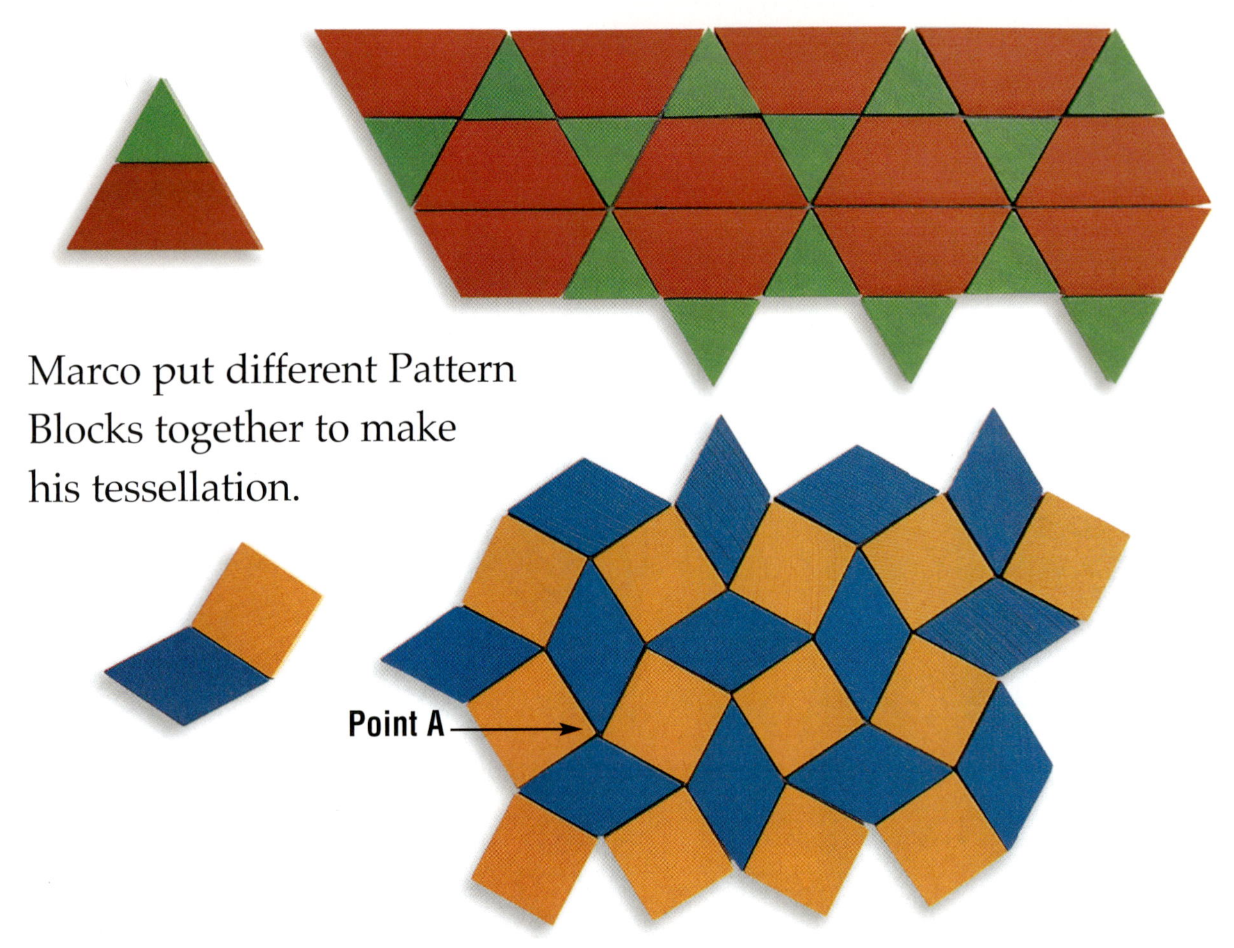

Marco put different Pattern Blocks together to make his tessellation.

2. How did Marco make his tessellation? Describe how he flipped and turned his new shape.

3. How many angles can you see at Point A? Which of the angles look the same? Which of the angles look different?

4. Work with a partner. Find 2 Pattern Blocks you think can be used to make a tessellation.
   Make a design to check whether you were correct.

Use these instructions to make a cube.

Step 1: Use coffee filters or cut out 6 circles, all the same size.

Step 2: Fold each circle in half to show its *diameter*. Unfold it.

Step 3: Fold the circle again.

Step 4: Draw lines to connect the ends of the 2 diameters.

Step 5: Repeat for all the other circles.

Step 6: To assemble the cube, fold along the sides of all the squares to make flaps. Staple or glue the flaps together to create the cube.

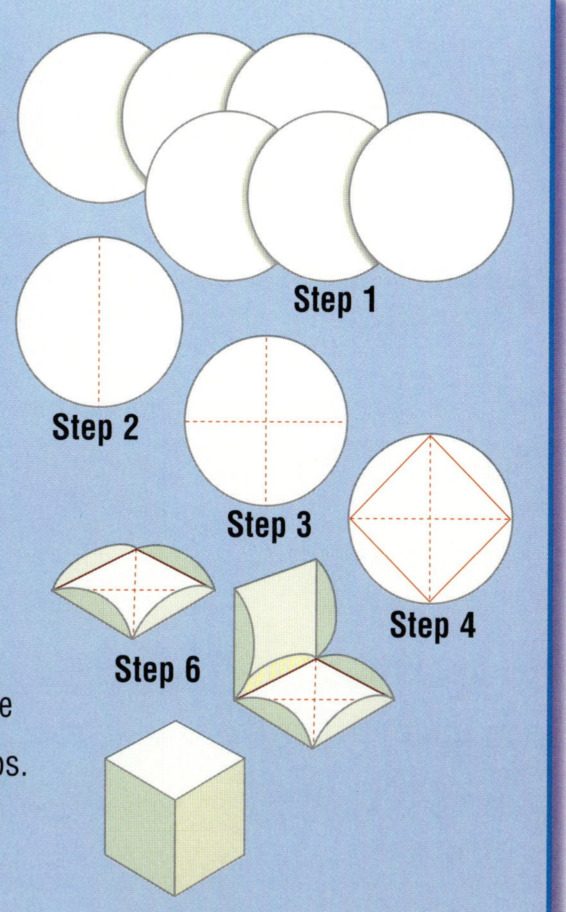

1. Look at the diagram. Describe the
   a. lines
   b. angles
   c. shapes
   What special names do you know?

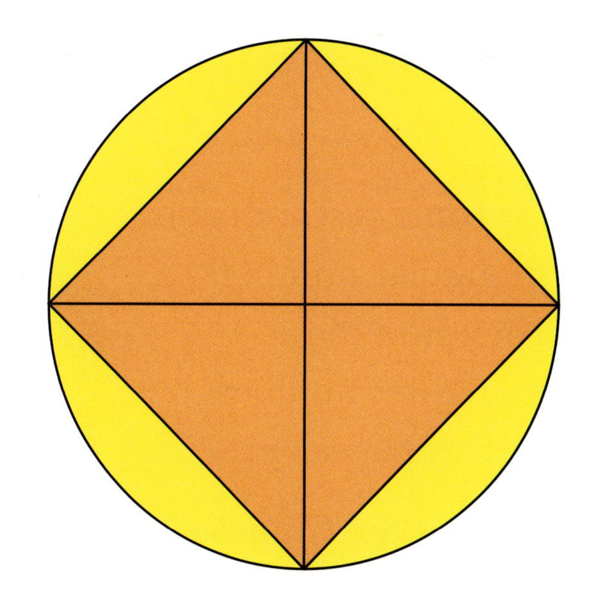

2. Find all the right angles in the drawing. Where do you see them?

3. Look at the angles at each corner of the square. What do you notice?

Some students placed points on circles. They drew lines to join the points. Lines that join two points are called *line segments.*

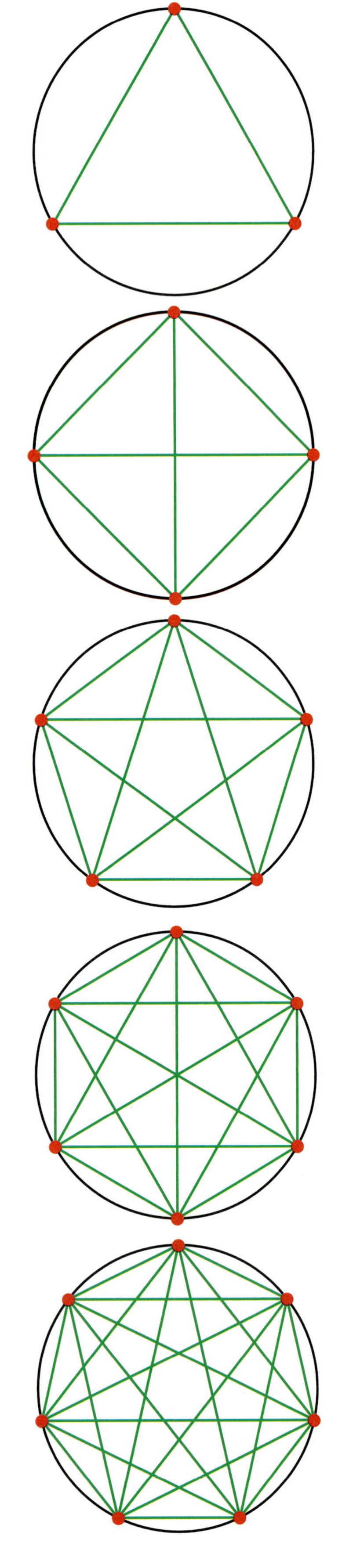

1. Count the number of points on each circle. What pattern do you see?
2. What shapes do you see within each circle?
3. Copy this table. Count the connecting lines on each circle. Write the missing numbers.

Number of points	Number of connecting lines
3	3
4	6
5	
6	
7	

4. What number of connecting lines do you predict you will get for circles with
   a. 8 points? **b.** 9 points?
5. How did you make your prediction?
6. Do some research about triangular numbers. Then find the triangular numbers on this page.

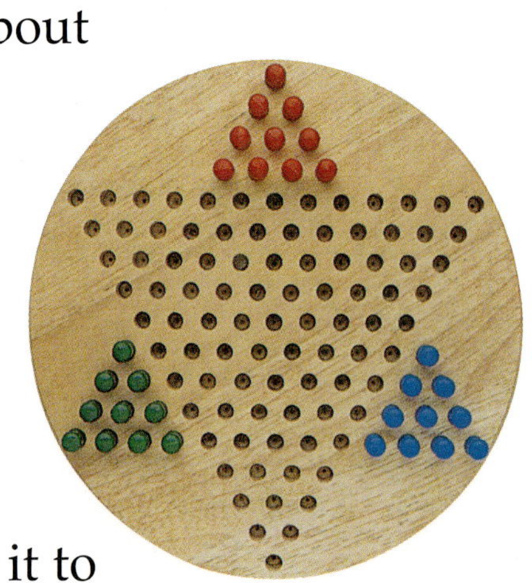

7. Look at the Chinese checkers board. How could you use it to show triangular numbers?

Look at these results from the Golden Springs athletics competition.

Name	Shot Put	Javelin	Discus
Dave	10.9 m	55.7 m	34.65 m
Rollo	11.1 m	59.6 m	32.15 m
Sal	8.2 m	56.95 m	32.5 m
Selina	8.55 m	57.5 m	33.8 m
Tim	12.85 m	51.25 m	33.08 m

1. Name the winners of the three events.
2. Name the competitors who came 2nd and 3rd in each event.
3. Calculate the difference between Dave's and Sal's results in the
   a. shot put **b.** discus **c.** javelin
4. Here are the records for each throwing event.

Shot Put	Javelin	Discus
13.6 m	59.04 m	35.2 m

   a. In which event was the record broken?
   b. By how much was the record broken?
5. How much short of the record were the other two winners' results? Explain how you figured out the answers.

1. Look at the whole square. How many tenths are in one whole?

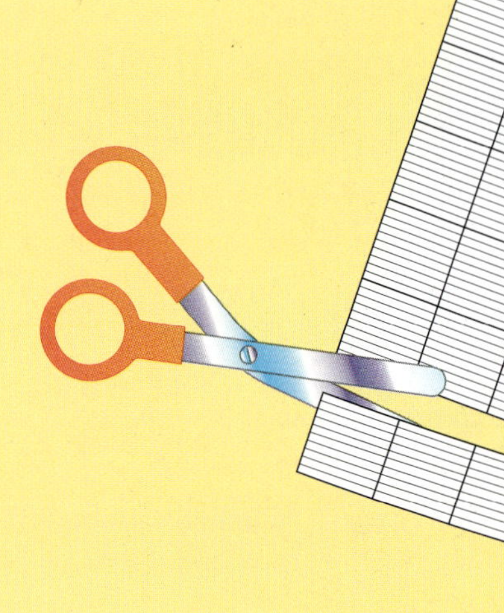

2. Delvina cut one-tenth from her whole square. How many more tenths could she cut out?

3. Delvina cut her *one-tenth* piece into *ten equal* pieces. Look at the piece she cut.
   a. How many of these pieces are in the whole square?
   b. What fraction name would you give to one of these pieces?

4. Next Delvina cut one of her *hundredth* pieces into *ten equal* pieces.
   a. How many of these equal pieces are in the whole square?
   b. What fraction name would you give to one of these pieces?

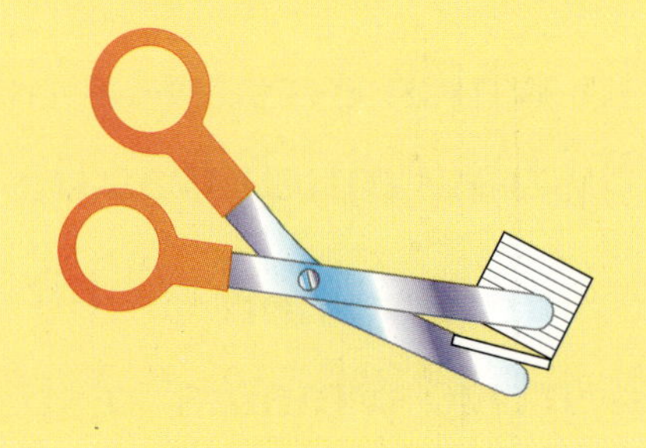

When you enter a number, the computer changes it.
It adds to, or subtracts from, one place only.

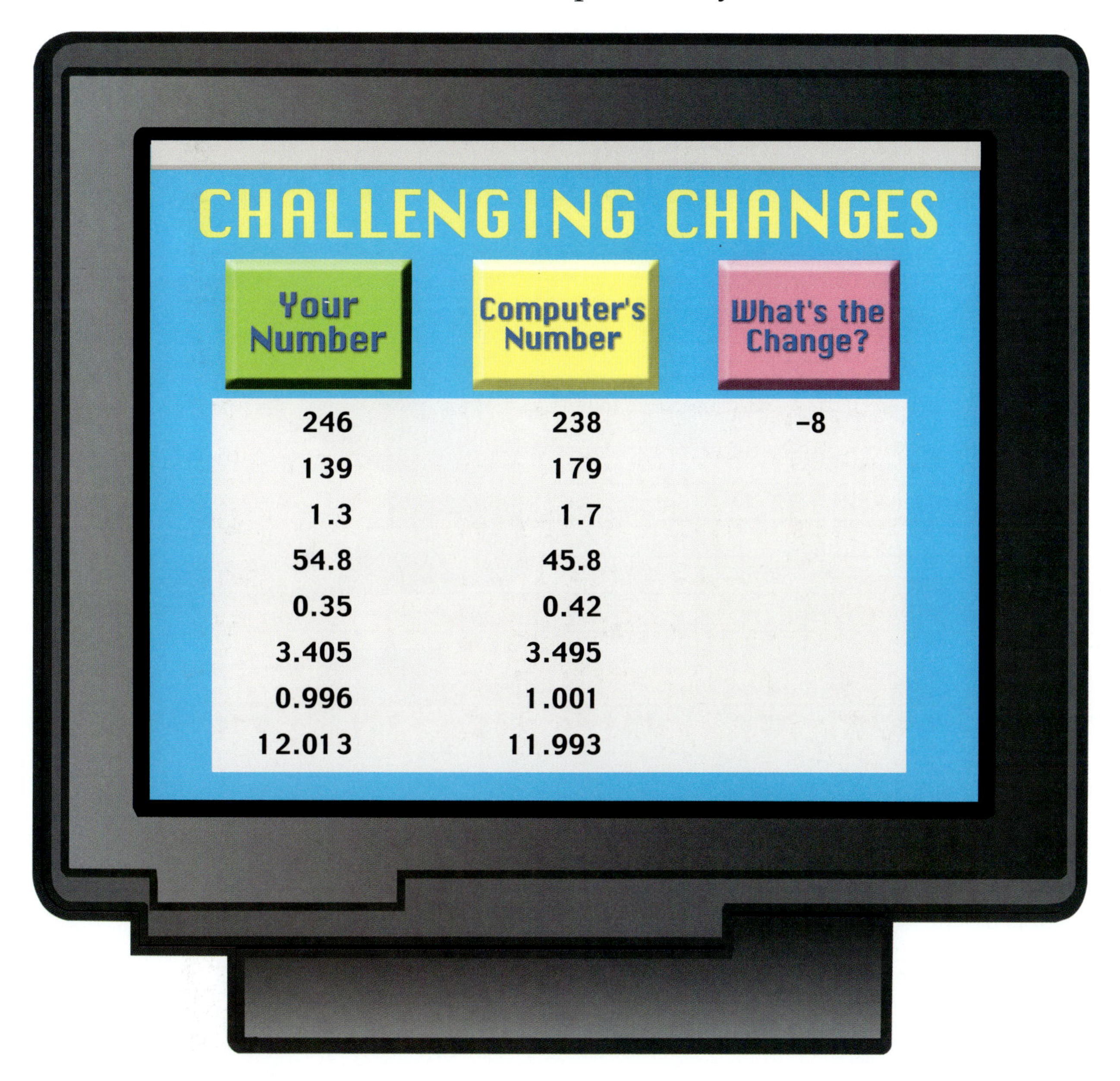
CHALLENGING CHANGES

Your Number	Computer's Number	What's the Change?
246	238	−8
139	179	
1.3	1.7	
54.8	45.8	
0.35	0.42	
3.405	3.495	
0.996	1.001	
12.013	11.993	

1. Look at each number.
   a. In which place did the computer make the change?
   b. What number was added or subtracted?
   c. Explain what happened when the change was made.
2. Which changes were easy to figure out?
   Why were they easy?

Emily and Nicole put the coins shown above in a container. They dropped out 2 coins, added the values of the 2 coins, and then put them back in the container. They did 30 trials and then made a graph of the outcomes.

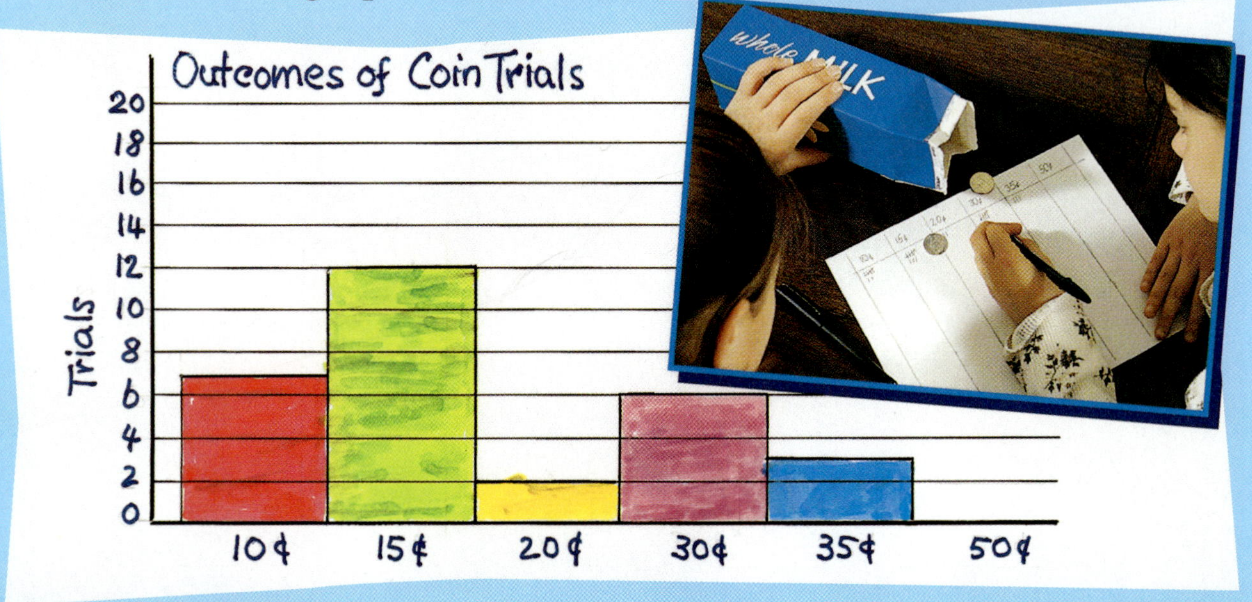

1. What outcomes
   a. did occur? b. did not occur?
2. What outcome occurred most frequently? Why do you think this happened?
3. Why do you think 50¢ was not an outcome?
4. If you made 60 trials, what do you think the number would be for each outcome?
5. Work with a partner. Put the 12 coins shown above in a container. Take turns dropping out 2 coins. Do 60 trials and then make a graph of the outcomes. Write a summary of what happened.

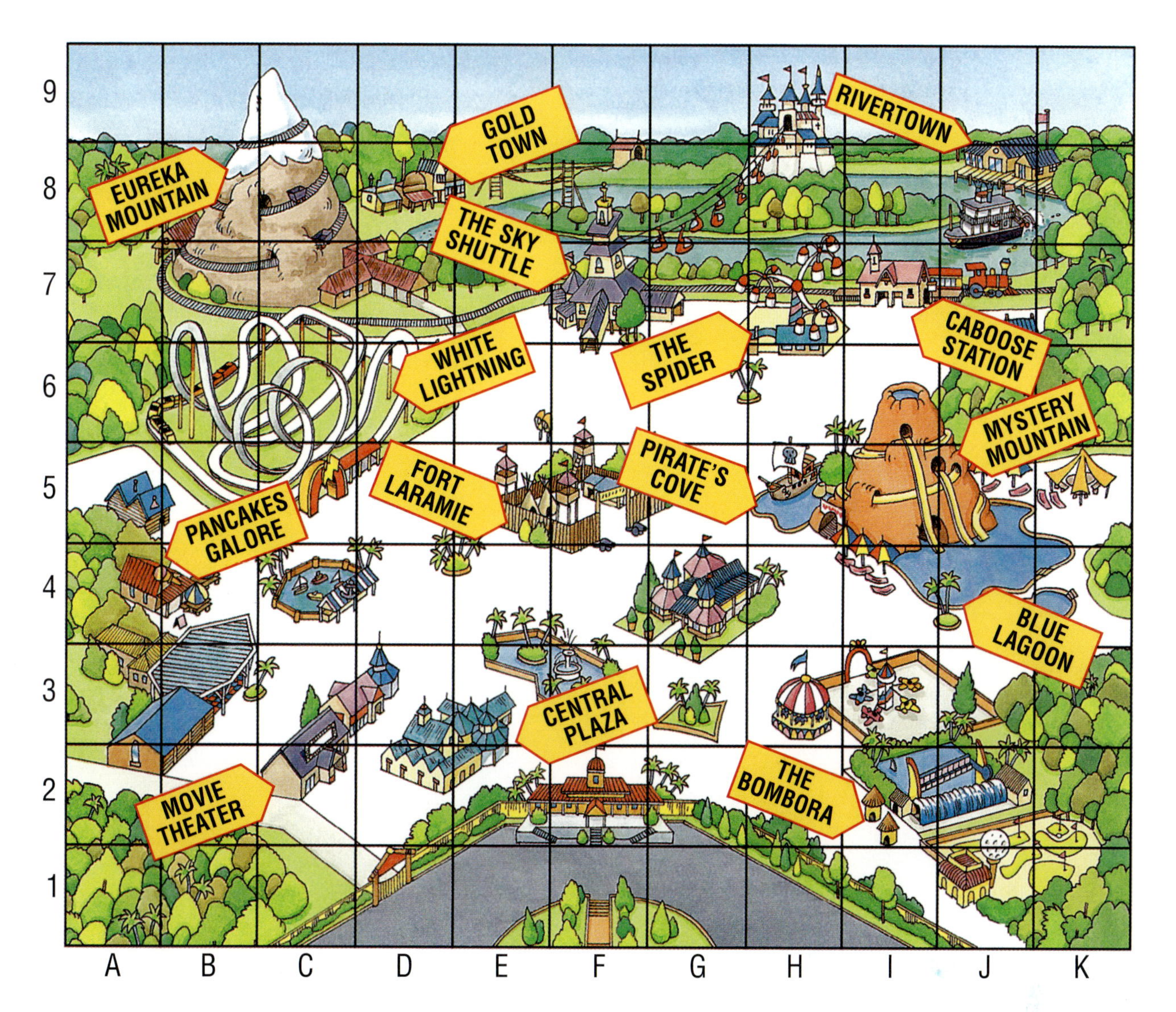

1. What information is given
   a. across the bottom of the map?
   b. along the side of the map?

2. Tell what you would find at each of these locations.
   a. D 6 b. F 7 c. J 5 d. B 8 e. E 5 f. I 7

3. What is the location of the entrance to each of these attractions?
   a. Rivertown b. Pirate's Cove c. Movie Theater

4. Plan a route to visit 5 attractions.
   Write the locations of the attractions in the order in which you will visit them.

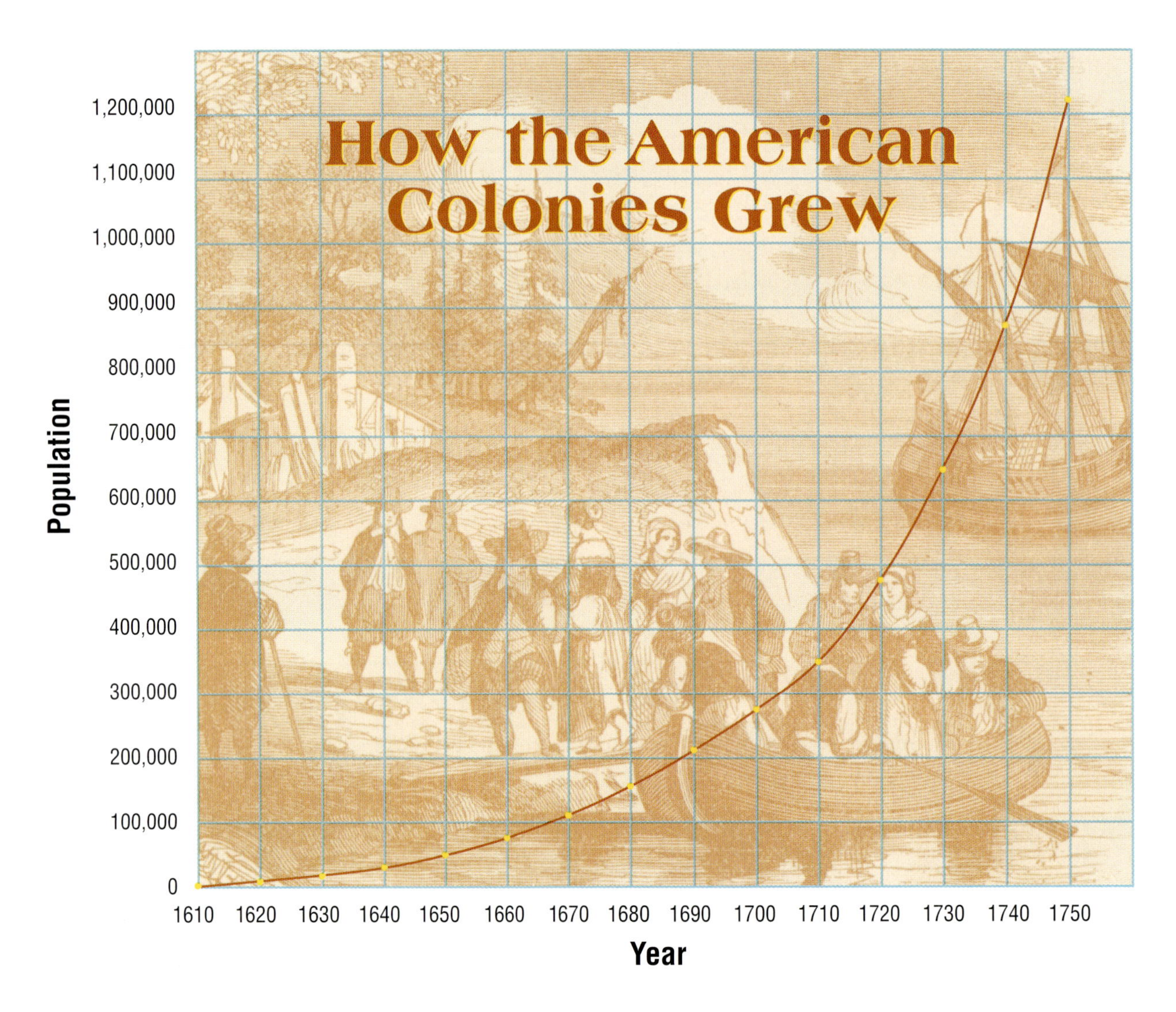

1. Look at the graph. What information is given along
   a. the horizontal axis? **b.** the vertical axis?

2. In what year was the population of the colonies about
   a. 100,000? **b.** 250,000? **c.** 500,000?

3. In which year do you think the population might have reached one million?

4. By how much did the population grow from
   a. 1660 to 1680? **b.** 1690 to 1710? **c.** 1710 to 1730?

5. Use the information shown to make a new line graph. Make the *vertical axis* go up in 50,000's to 500,000.

Sofia investigated the length of video titles. First she made a record of 100 titles. Then she made a circle graph to show the information.

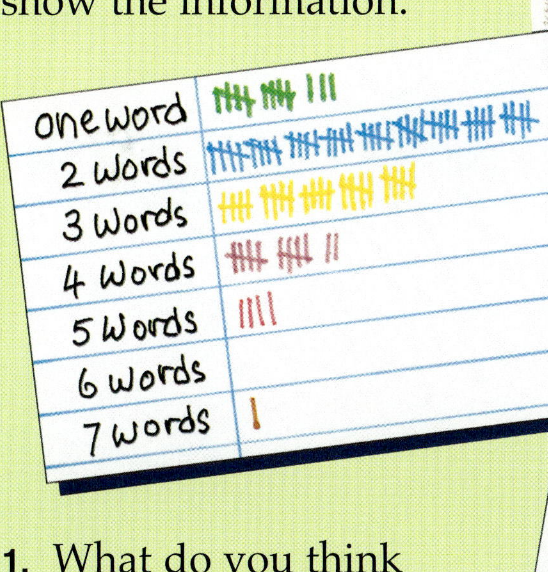

Circle graphs are sometimes called pie graphs.

1. What do you think Sofia did? Describe her circle graph.
2. What information can you see at a glance on Sofia's tally chart?
3. What is the circle graph better at showing?
4. What was the length of nearly half of the video titles?
5. What was the length of one-fourth of the video titles?
6. Work with a partner. Find a list of video titles. Record the length of 100 titles and construct a graph to show what you discovered.
7. Compare the information on your graph with that on Sofia's graph. Write a summary.

Wendy and Warren are helping arrange chairs.

1. List all the different ways 12 chairs can be put in equal rows.
2. List all the different ways 24 chairs can be put in equal rows. Compare the two lists. What do you notice?
3. List all the different ways 36 chairs can be put in equal rows. What do you notice?
4. List all the different ways 48 chairs can be put in equal rows. How do the lists for 24 and 36 chairs help you?
5. List the factors of each number in the following sets. Use the factors of the first number to help you find the factors of the other 2 numbers.

28, 56, 84

45, 90, 135

1. Look at the middle shelf. Describe how each type of food package has been arranged.
2. To find out how many of each type of food packages are on the shelf, what do you need to know?
3. The bread mix packages are stacked 6 deep. How many bread mix packages are on the shelf? How do you know?
4. How are these packages arranged?
   a. 24 boxes of Crunchy Crackers
   b. 30 boxes of Healthy Snacks
5. There are 36 cans of Simple Soup on the shelf. How are the cans arranged? How do you know?
6. How would you rearrange the 36 cans of Simple Soup to make the display only 3 cans wide?

Mr. Brian's class was planning a trip to the Eureka historic town. Each group in the class colored a grid to find the admission cost for 35 students.

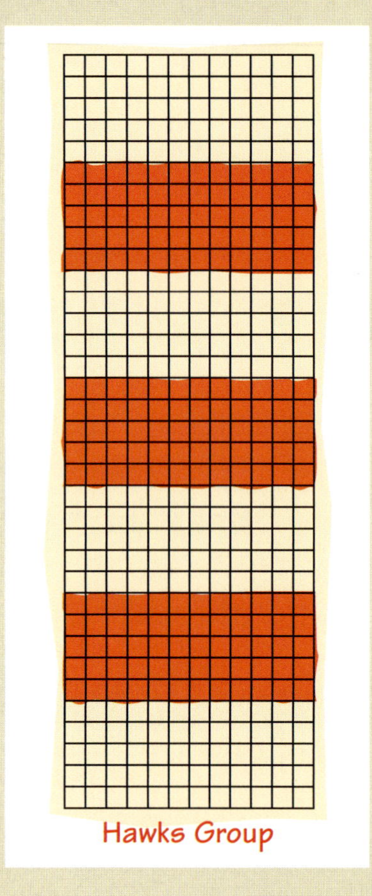

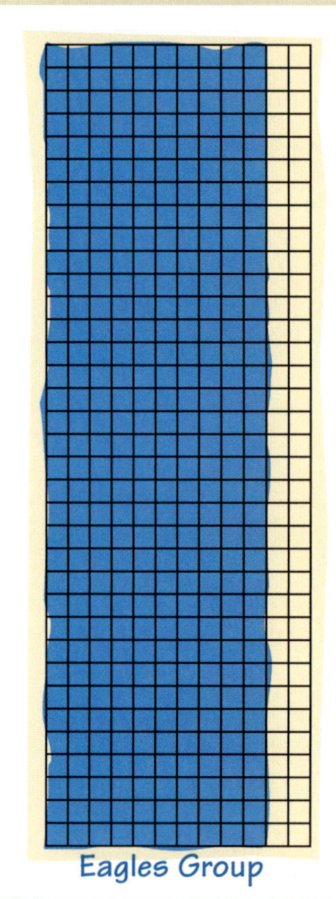

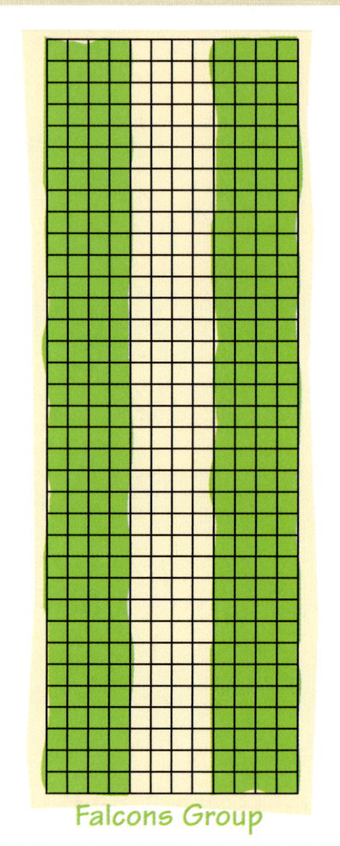

1. Describe how each group colored their grid.
2. How do you think each group figured out the total?
3. Work with a partner.
   - **a.** Cut out your own grid that shows 35×12. Try to find a different way to color it to find the cost of $35 \times \$12$.
   - **b.** Cut out grids to show these arrays. 15×36 25×28
   - **c.** Find a quick way to figure out the number of squares in each array.

1. Would 30 CD Doubles cost more or less than $600? How do you know?

2. Which of these purchases would cost more than $500? Explain how you know.
   a. 20 CD Boxed Sets
   b. 25 CD Doubles
   c. 55 Old Favorites
   d. 32 CDs at $16 each

   Which estimates could you make easily in your head? Why?

3. On Friday, the CD Cellar had a sales goal of $800. Which of these would achieve the sales goal?

   64 CDs at $12

   or

   45 CDs at $18

   How did you decide?

4. Find some other ways the CD Cellar could make their sales goal of $800.

Some students each drew a rectangle on grid paper. Then they drew a triangle on their rectangle.

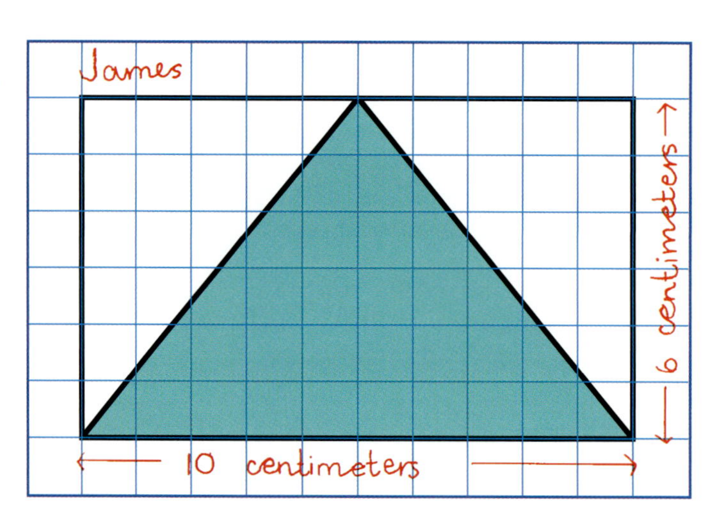

1. Look at each rectangle. What is its height? What is its width?
2. Look at each triangle. What is its height? What is the width of its base? What do you notice?

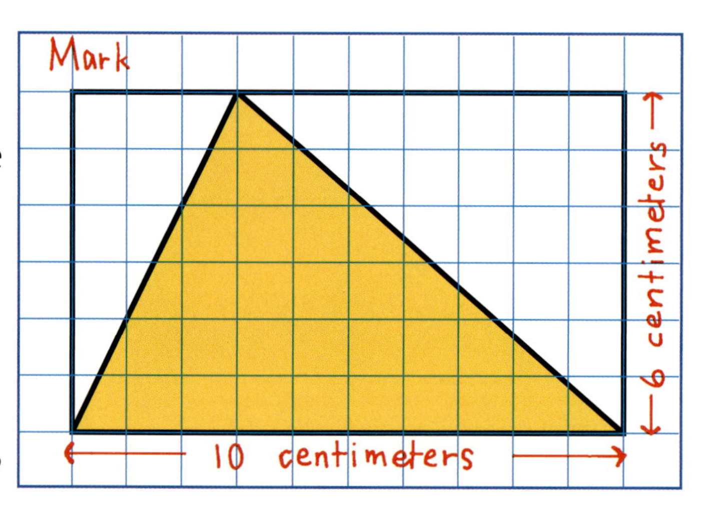

3. Make an estimate of the area of James's triangle. How did you do it?
4. Now estimate the area of the other 2 triangles. What did you discover?
5. Work with a partner to find the area of some other triangles. Fold or cut the triangles to save time.

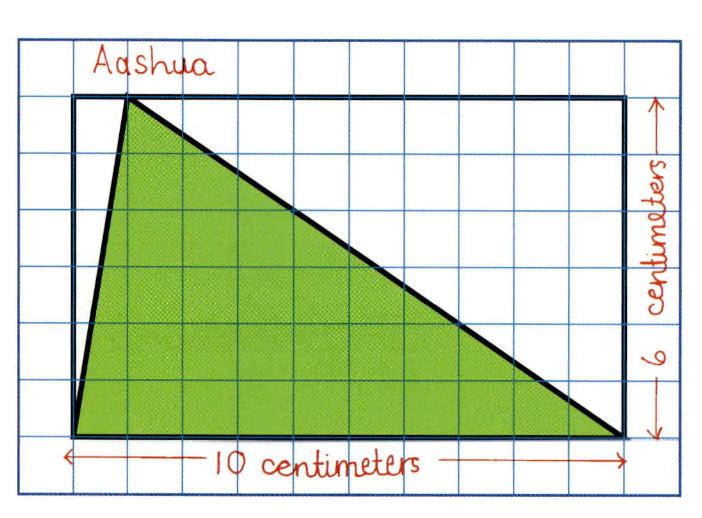

Mr. Sullivan made a sketch of his patio. He will use it to find out how many tiles he needs.

1. How many 1-foot square tiles does Mr. Sullivan need to buy? How do you know?

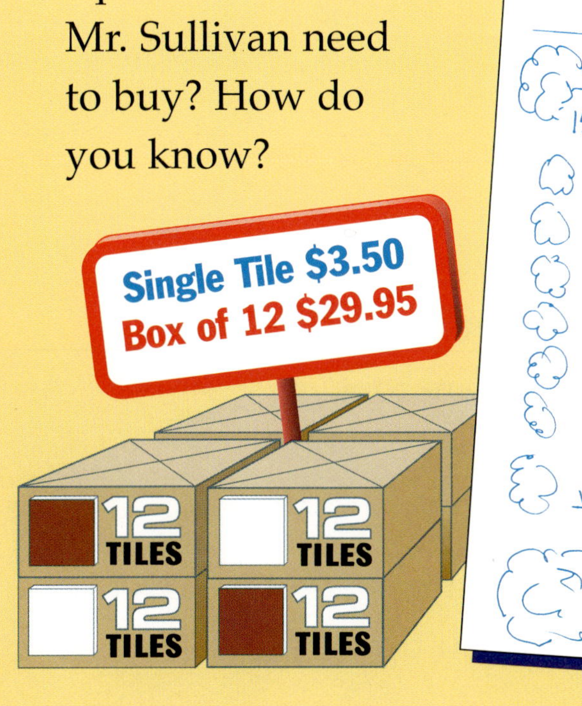

2. What is the best way for Mr. Sullivan to buy his tiles? What will his tiles cost? How did you figure it out?

3. Suppose Mr. Sullivan decided to use brown tiles around the edge and white tiles for the rest of the patio. Sketch his design on grid paper.

4. How many tiles of each color will Mr. Sullivan need to buy? How did you figure it out?

5. What is the best way for Mr. Sullivan to buy the tiles for his new design? What will the cost be?

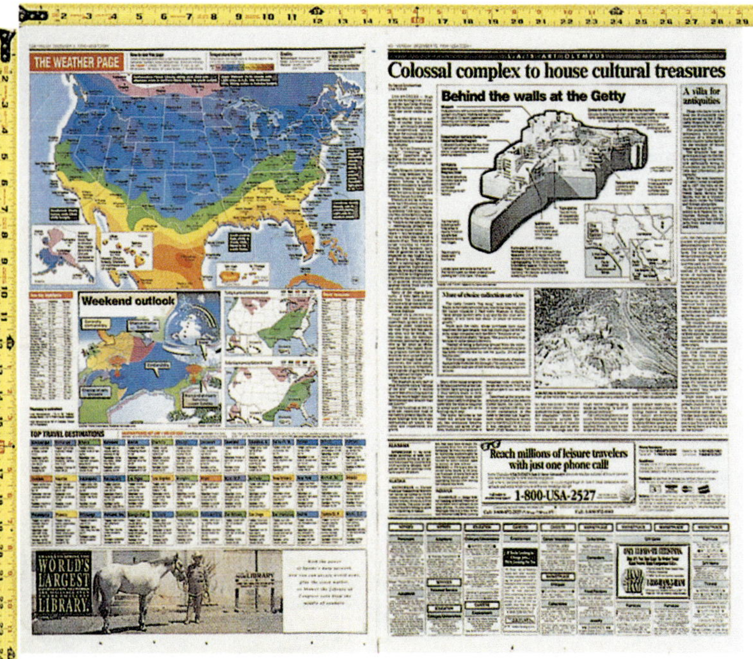

1. Look at this sheet of newspaper. Is it more or less than
   a. 1 square foot?
   b. 1 square yard?

2. Use some pages from your local newspaper.
   a. Overlap the pages to show 1 square yard.
   b. Cut out some pieces that are 1 square foot.
   c. Find out how many 1 square foot pieces you need to make 1 square yard.

3. How many square yards of linoleum are needed to cover each of these floor areas? How did you figure it out?

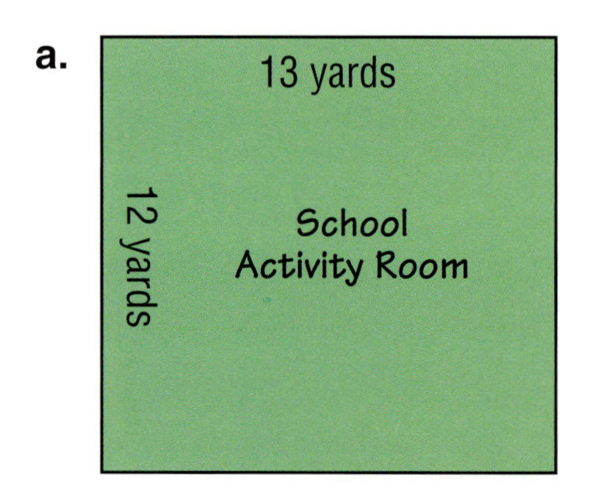

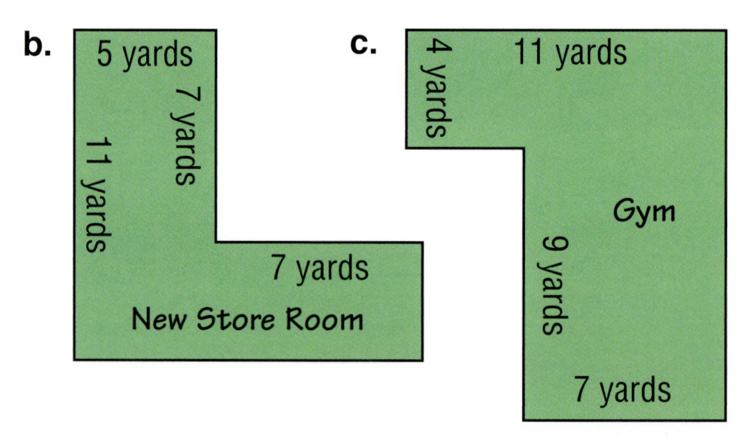

4. Use grid paper to draw floor plans that show 100 square yards. Try to draw plans that are *not* rectangles.

The Sunnydale Recreation Committee wanted to repaint the surface of their 3 playing courts.

1. How do you think the committee would figure out the amount of paint to buy? (All courts are painted the same color.)
2. Look at the dimensions of the courts.
   - **a.** Which court will need the *least* paint?
   - **b.** Which court will need the *most* paint?
3. What is the total area of all 3 courts? Explain how you found the answer.
4. About how many gallons of paint will the committee need to paint the 3 courts? Explain how you made your estimate.
5. About how much will the paint cost? How do you know?

Alex, Sun, and Dana were training for a marathon. They figured out how far each person would need to run each week.

Weekly Training Schedule

Week 1	120 km
Week 2	122 km
Week 3	124 km
Week 4	126 km
Week 5	128 km
Week 6	130 km

Weekly Training Plan

Alex	–	4 days
Sun	–	5 days
Dana	–	6 days

1. All runners had to run the same distance, but they each had a different number of days for training. How far did the runners have to run each training day in Week 1?
   a. Alex **b.** Sun **c.** Dana
2. In what way did each runner's daily training distance change from week to week?
3. Copy and complete the chart. What patterns do you notice?

Distance run per training day

	Alex	Sun	Dana
Week 1			
Week 2			
Week 3			
Week 4			
Week 5			

These students are figuring out prices for items sold in their store.

1. This is the way the school buys pens.

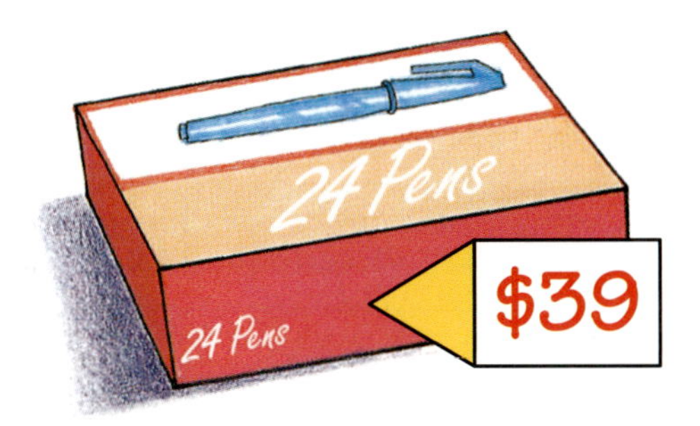

What do you think the students will charge for pens sold in packs of

a. 6?

b. 4?

c. 3?

d. 2?

2. Explain how you made your decisions in Question 1. How did you use division? How did you use multiplication?

3. What are some ways the students could break up each of the packs below? What price do you think the students will charge for each small pack they make?

a.

b.

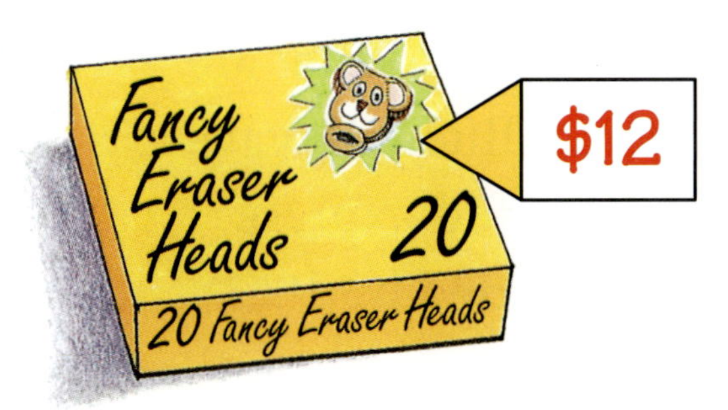

Each Sunday, *The Globe* newspaper is delivered to vending machines around the town. Ten *Globes* are placed in each vending machine.

SUNDAY GLOBE DELIVERIES					
VANS MAKING DELIVERIES	VAN 1	VAN 2	VAN 3	VAN 4	VAN 5
NUMBER OF NEWSPAPERS TO DELIVER	380	440	670	580	520

1. How many vending machines does each van driver fill?
   a. Van 1 **b.** Van 2 **c.** Van 3 **d.** Van 4 **e.** Van 5
2. What is the total number of *Globes* delivered on Sundays?
3. To how many vending machines are *Globes* delivered each Sunday? What are some different ways to find the answer?
4. Explain what happened when you divided the number of newspapers by 10.
5. Divide these amounts by 10.
   a. $875
   b. 225 feet of chain
   c. 475 newspapers
   What did you do with the 5 that was left over each time? What are some different ways you could express "5 divided by 10"?

Some fourth-grade classes collected all the stamps their families received each week. They did this for 5 weeks and then made a graph to show how many stamps they had collected.

Week beginning	Stamps collected over 5 weeks (one square = 50)
February 4	
February 11	
February 18	
February 25	
March 4	

1. In which week was the greatest number of stamps collected? How does this compare with each of the other weeks?
2. Why do you think so many more stamps were collected in one week than the other weeks?
3. How many stamps were collected during the 5-week period?
4. Look at the graph. About how many stamps were collected each week? How did you make your estimate?
5. Find the *mean* for the number of stamps collected each week during the 5-week period.

> If you divide the total number of stamps by the number of weeks, you find the *mean*. The mean is an *average*.

1. What was the length of the table in
   a. centimeters? b. millimeters?
   How do you know?

2. If the table was a little *less* than $1\frac{1}{4}$ meters, which of these could be its length?
   a. 1.15 meters b. 1.095 meters c. 1.605 meters
   d. 1.275 meters e. 1.3 meters f. 1.2 meters

3. If the table was a little *longer* than $1\frac{1}{4}$ meters, which of these could be its length?
   a. 1.22 meters b. 1.27 meters c. 1.72 meters
   d. 1.209 meters e. 1.218 meters f. 1.255 meters

4. Delorte found that the table was $\frac{3}{4}$ meter high. What was its height in
   a. tenths of a meter? b. hundredths of a meter?
   c. thousandths of a meter?

5. The table was $\frac{1}{2}$ meter wide. What was its width in
   a. tenths of a meter? b. hundredths of a meter?
   c. thousandths of a meter?

BUILDING COPPER TRAILS

Students at Dogwood Elementary School saved pennies for a whole year. Their goal was to lay a 1-kilometer trail of pennies.

1. Imagine you measured the trail when it was each of these lengths. Which length is closest to

   300 meters
   456 meters
   810 meters

   a. 1 kilometer? **b.** $\frac{1}{2}$ kilometer? **c.** $\frac{1}{4}$ kilometer?

2. The fourth-grade classes at Dogwood laid out 4 separate trails of the pennies they had collected.

Pennies collected by fourth-grade students			
Nov. 1	Jan. 1	March 1	May 1
0.163 km	0.121 km	0.089 km	0.187 km

   a. On January 1, was the total length of their first 2 trails closer to $\frac{1}{2}$ kilometer or $\frac{1}{4}$ kilometer?

   b. On May 1, was the total length of their 4 trails closer to 1 kilometer or $\frac{1}{2}$ kilometer? How do you know?

3. The students in the whole school laid out 4 separate trails of the pennies they had collected.

Pennies collected by all students			
Nov. 1	Jan. 1	March 1	May 1
0.472 km	0.201 km	0.384 km	0.618 km

   a. Which 2 trails had a total length of about 1 kilometer?

   b. Was the total length of the 4 trails about 1 kilometer, $1\frac{1}{2}$ kilometers, or 2 kilometers?

4. How successful were Dogwood Elementary students in achieving their goal?

A fourth-grade class observed water in a jar over a 5-day period.

1. Describe what the graph shows.
2. What happened to the level of the water over the 5-day period?
3. When did the water level reach
   a. 6.5 centimeters? **b.** 5.1 centimeters?
4. By what amount did the water level fall between
   a. Day 1 and Day 2? **b.** Day 4 and Day 5?
5. Between which 2 days was the greatest fall in water level?
6. What was the total decrease in the water level for the 5 days?
7. To what level do you think the water would fall on Day 6? Explain why you made that prediction.

1. Would $5 be enough to buy 2 of these packages of
   a. ham? **b.** pastrami?

2. Would it cost more or less than $10 to buy
   a. 3 packages of ham? **b.** 2 packages of chicken?
   c. 2 packages of turkey? **d.** 4 packages of pastrami?

3. Which items in Question 2 have a total weight of about 2 pounds? How do you know?

4. A recipe requires 2 pounds of chicken. How much would the chicken cost? How did you figure it out?

5. A sandwich requires $\frac{1}{4}$ pound of meat. How many sandwiches could be made from one package of pastrami? About how much would the meat in each sandwich cost?

You will need cubes.

1. How many cubes were used to make this box shape? How can you tell?

2. Work with a partner. Count out 24 cubes.
   a. Copy the box shape in the picture.
   b. Use 24 cubes to build some other box shapes.
   c. Describe how your box shapes are alike. How are they different?

3. Build as many box shapes as possible using:
   a. 18 cubes b. 36 cubes

4. Which 2 box shapes below have exactly the same number of cubes? How are those 2 box shapes alike? How are they different?

1. Work with a partner.
   Use centimeter grid paper to make a box like the one below.

2. How many 1-centimeter cubes do you think you would need to fill your box? Give reasons for your answer.

3. Fill the box you made. Count the cubes.

4. Look at these box plans.
   a. Which box do you think will hold more?
   b. Use centimeter grid paper to make boxes like the ones shown.
   c. Fill your boxes with centimeter cubes. Was your prediction correct?

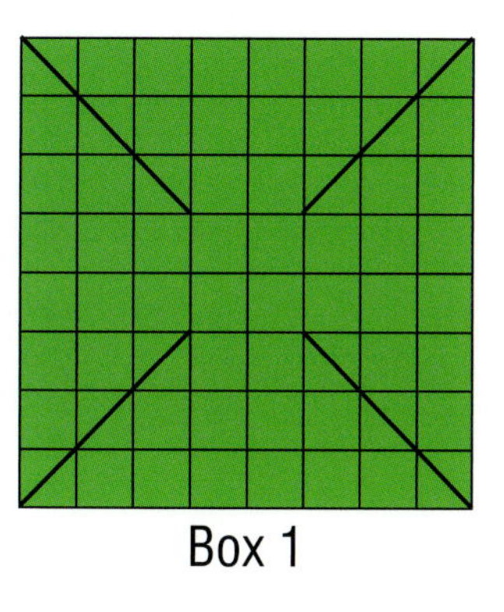

Box 1

Box 2

5. Compare the size of a 2-centimeter cube and a 1-centimeter cube. How many 2-centimeter cubes do you think you need to fill each box in Question 4? Check your prediction.

1. How many batches of plain pancakes could you make with
   a. 1 quart of milk?
   b. $\frac{1}{2}$ gallon of milk?
2. What will you do to the recipe if you want to make
   a. 16 plain pancakes?
   b. 40 plain pancakes?
3. What quantity of flour and milk is needed to make
   a. 16 plain pancakes?
   b. 40 plain pancakes?

Plain Pancakes

$1\frac{1}{4}$ cups flour
1 cup milk
1 egg
3 teaspoons baking powder

Makes 8 pancakes

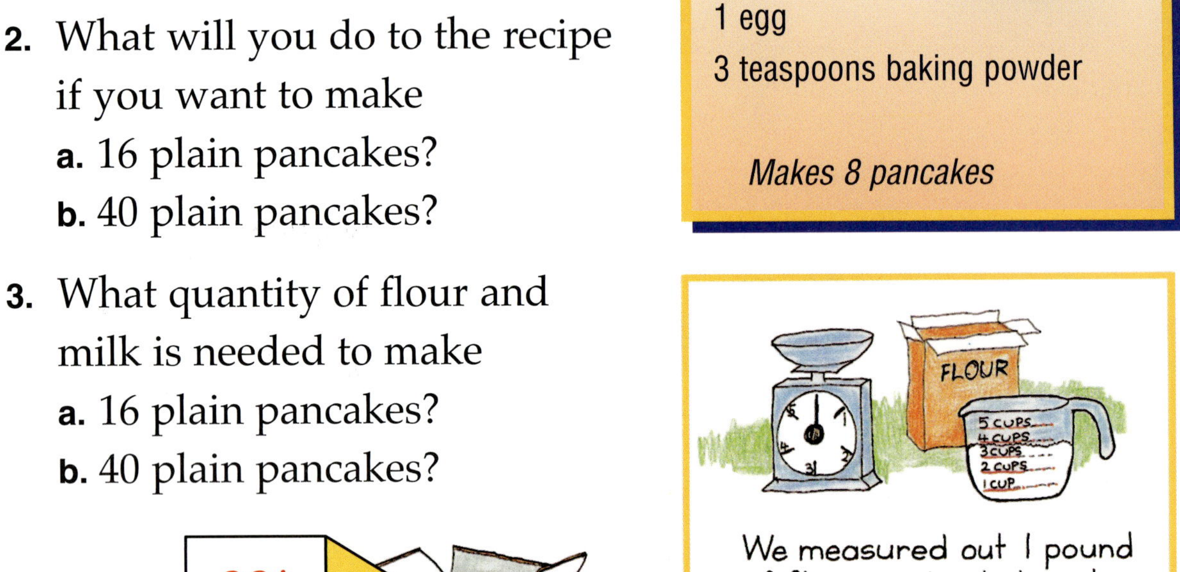

4. How much flour and milk do you need to make 60 apple pancakes?
5. Estimate how much it would cost to buy the main ingredients for 60 apple pancakes.

Coral and Alexi experimented with pouring 1 liter of juice into smaller containers. They found that they could pour 1 liter into ten 100-milliliter containers.

Milliliter Facts

A milliliter is the unit used in hospitals to measure medicines and other liquids. It is also the unit used in scientific experiments. The abbreviation for milliliter is **mL**. There are 1,000 milliliters in 1 liter.

1. How many milliliters are there in 1 liter?
2. How many 10-milliliter containers could Coral and Alexi fill with 1 liter of juice?
3. Look at the 2 pictures. Which measurer is holding 2 mL? Which measurer is holding 6 mL?

a.

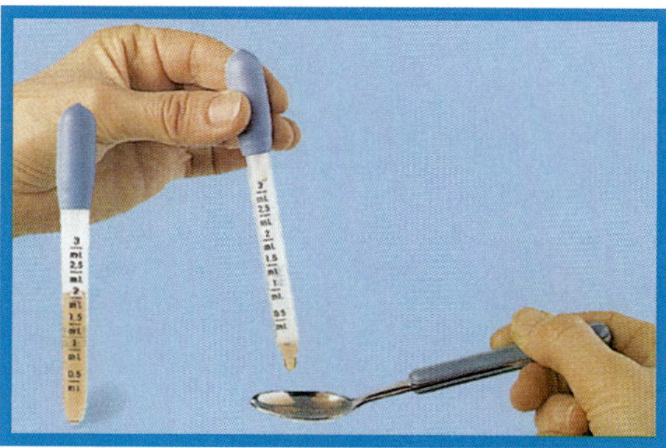

b.

4. About how much do you think this lid holds?

5 mL | 10 mL | 100 mL

These students were investigating the weight of 1 liter of water. They discovered that 1 liter of water weighs exactly 1 kilogram.

1. What do you remember about the kilogram? How many grams are in 1 kilogram?
2. What do you remember about the liter? How many milliliters are in 1 liter?
3. If 1,000 milliliters of water weigh 1,000 grams, how many grams are there in
   a. 1 liter?
   b. $\frac{1}{2}$ liter?
   c. 100 milliliters?
   d. 10 milliliters?

The students were learning about healthy food. They found out about fat.

4. Work in groups.
   a. Weigh some foods to see what 40 grams looks like.
   b. Fill a small container with 40 milliliters of water. What do you think it will weigh? Check your prediction.

1. What is the total amount of fat in each of these meals?

a.

b.

2. What is the total amount of fat for the 2 meals in Question 1?

3. Choose foods from the list to plan dinner for yourself. Find the amount of fat in your dinner.

4. Choose food for yourself for one day. Try to keep the total amount of fat less than 40 grams.

Grams of Fat in Foods

Breads and Cereals		***Drinks***	
Bread (slice)	1.0 g	Fruit juices	0 g
Cracker (one)	5.5 g	Soft drinks	0 g
Cupcake	6.0 g	***Fruits***	
Doughnut	19.0 g	All fresh fruit	0 g
Muffin	6.0 g	***Meat/Fish/Eggs***	
Spaghetti (serve)	0.5 g	Egg (boiled)	5 g
Rice (serve)	0.5 g	Fish	2.5 g–5 g
Corn Flakes	0.5 g	Hamburger	18 g
Oats	2.5 g	Chicken (no skin)	6 g
Dairy		Beef steak	8 g–11g
Cheese		***Snack Foods***	
Cheddar	6.5 g	Popcorn (no oil)	2.5 g
Cream	7.5 g	Nuts	7 g–11 g
Mozzarella	4.5 g	Potato Chips	8 g–13 g
Fruit Yogurt	5.5 g	***Vegetables***	
Ice Cream	5.0 g	(fresh or	
Milk (whole)	9.5 g	steamed)	0 g
Milk (skim)	0 g		

- All amounts are average portions.
- If cooking in butter or margarine, add 4 g.

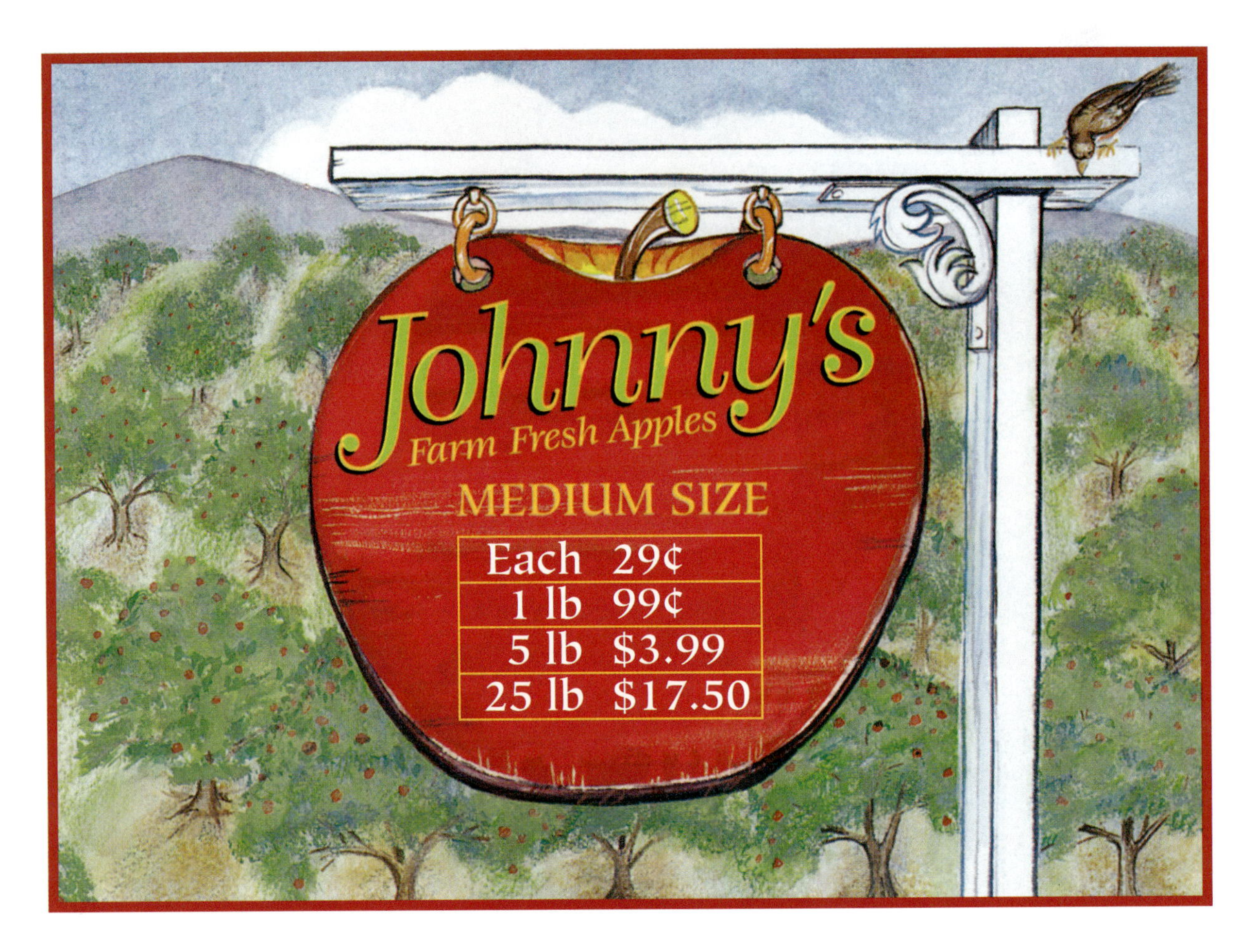

1. Estimate the number of medium-size apples in a:
   a. 1-lb bag
   b. 5-lb bag
   c. 25-lb box

   How did you make your estimates?

A medium-size apple weighs about 4 ounces.

2. Compare the price of a single apple with the price of an apple bought in a 1-pound bag. Explain your answer.

3. Make some other comparisons using the weights and prices shown on Johnny's sign.

1. Which animal in the list has the closest weight to 10,000 pounds?
2. What is the difference in weight between the African and the Indian elephants?
3. Which two animals have the least difference in weight?
4. One ton is 2,000 pounds. Which animal weighs the closest to 1 ton?
5. Which animal weighs the closest to 2 tons? How do you know?
6. How much more or less than 1 ton do these animals weigh?
   a. bison   b. giraffe
7. About how many tons do these animals weigh?
   a. hippopotamus
   b. Indian elephant
   c. African elephant

These are large specimens, but not the largest on record. Source: *The Top 10 of Everything.*

Heaviest Land Animals

	pounds
African elephant	11,150
Indian elephant	8,875
White rhinoceros	4,850
Hippopotamus	4,400
Giraffe	2,625
Crocodile	2,425
Asian gaur	1,975
Bison	1,825
Kodiak bear	1,800
Alaskan moose	1,750

Ed and Deedee experimented with long pieces of cord. They each tied the ends with a knot and then made "over and under" paths.

1. Take turns following the paths Ed and Deedee made.
   a. Start at the knot and use your finger to follow the path back to the knot. Do not go over any part twice.
   b. Do it again, this time moving in the other direction.
2. Look at each pipe cleaner design below. For each design:
   a. Start at any point and try to follow the path without going over any part twice.
   b. What did you discover?

3. Work with a partner and try to copy each design.
4. Draw or make your own "over and under" paths. Challenge other members of the class to follow your paths.

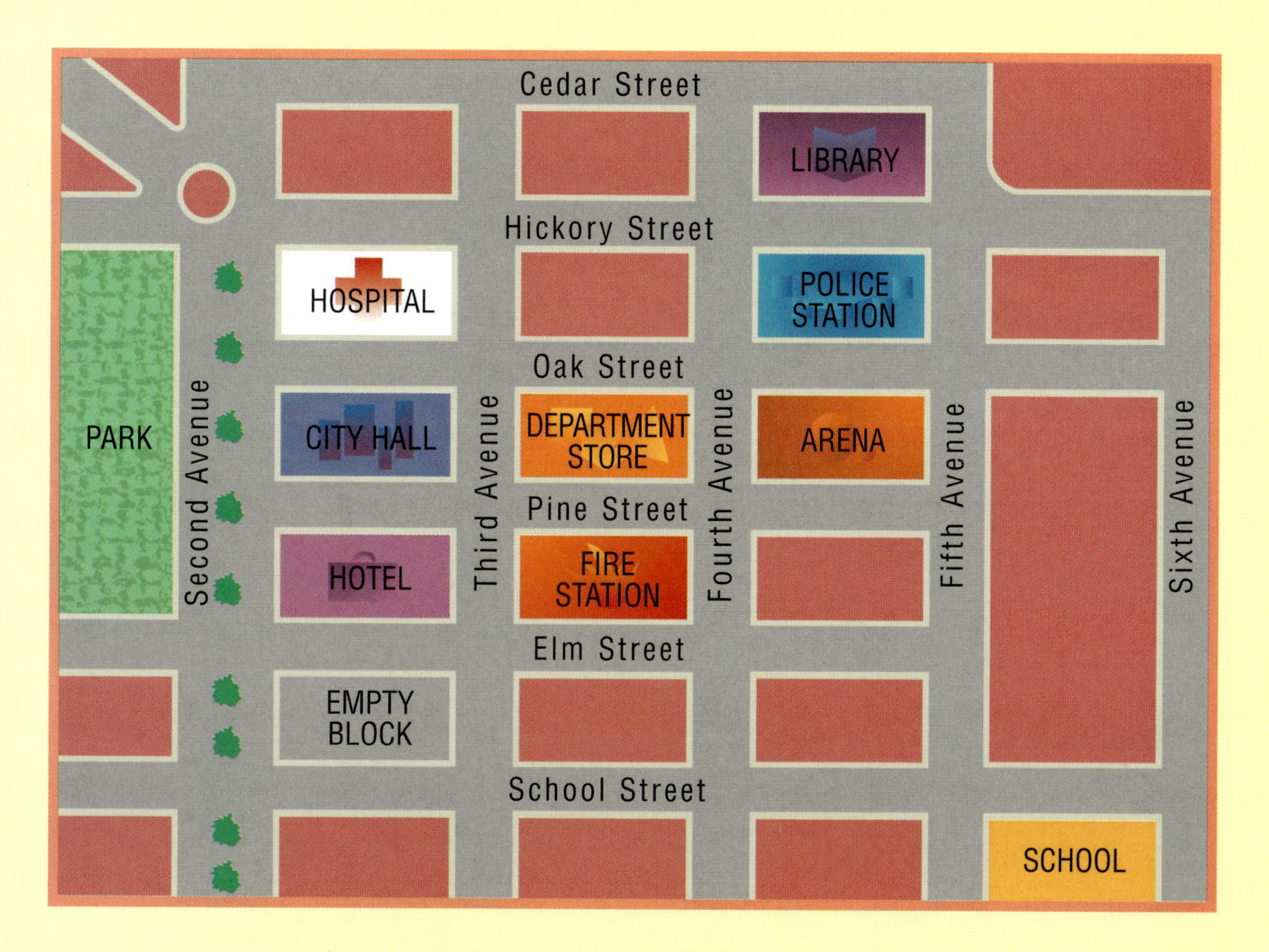

1. If you were at the fire station, what route would you take to get to school?
2. What streets would you walk along if you took *either* of the two shortest routes from the fire station to the school?
3. Suppose you are at the hotel. What is *one* of the shortest routes you could take to school? What is another way that is just as short? How many different "shortest routes" are there?
4. Find *all* the "shortest routes" from the department store to the school. What did you discover?
5. How many "shortest routes" are there from City Hall to the school? How did you figure out the answer?
6. What pattern do you notice about the shortest routes?

Look at the Pattern Block house shapes that some students made.

1. Look at each house shape.
   a. What shapes do you see?
   b. How do the house shapes change each time?
   c. How many Pattern Blocks were used to make each house shape? How did you figure it out?
2. How many Pattern Blocks do you predict will be needed for the next house shape? How did you make your prediction?

3. What is the length of each side of each Pattern Block?
4. Find the perimeter of each house shape. What did you notice?
5. What do you think the perimeter of the next house shape will be? How did you make your prediction?